How to
Write and Publish
a Scientific Paper

The Professional Writing Series

This volume is one of a series published by ISI Press®. The series is designed to improve the communication skills of professional men and women, as well as students embarking upon professional careers.

Books published in this series:

How to
Write and Publish
a Scientific Paper

Second Edition

ROBERT A. DAY

Philadelphia

Published by

ISi PRESS® A Subsidiary of the
Institute for Scientific Information®
3501 Market St., University City Science Center, Philadelphia, PA 19104 U.S.A.

© 1979, 1983 Robert A. Day

Library of Congress Cataloging in Publication Data

Day, Robert A., 1924–
How to write and publish a scientific paper.

(The Professional Writing Series)
Bibliography: p.
Includes index.
1. Technical writing. I. Title. II. Series.
T11.D33   808′.0665021   83-8460
ISBN 0-89495-021-5
ISBN 0-89495-022-3 (pbk)

Printed in the United States of America
90 89 88 87 86 85 84 83   8 7 6 5 4 3 2 1

Contents

Preface

*The Man of Science appears to be the only
man who has something to say just now—and the
only man who does not know how to say it.*
                                        —SIR JAMES BARRIE

There are four things that make this world go round: love, energy, materials, and information. We see about us a critical shortage of the first commodity, a near-critical shortage of the second, increasing shortage of the third, but an absolute glut of the fourth.

We in science, of necessity, must contribute to the glut. But let us do it with love, especially love of the English language, which is the cornerstone of our intellectual heritage; let us also do it with energy, the energy we need to put into the scientific paper so that the reader will not need to use much energy to get the information out of the paper, and let us husband our materials, especially our words, so that we do not waste inordinate quantities of paper and ink in trying to tell the reader more than we know or more than the reader wants to know.

Whether or not one wholly subscribes to the "publish or perish" adage, there is no question but that the goal of scientific research is publication. Scientists, starting as graduate students, are measured primarily not by their dexterity in laboratory manipulation, not by their innate knowledge of either broad or narrow scientific subjects, and certainly not by their wit or charm; they are measured, and become known (or remain unknown), by their *publications*.

ix

A scientific experiment, no matter how spectacular the results, is not completed until the results are *published*. In fact, the cornerstone of the philosophy of science is based on the fundamental assumption that original research *must* be published; only thus can new scientific knowledge be authenticated and then added to the existing data base that we call science. This concept was given added weight when it was accepted as a matter of national policy. Current government policy, first proclaimed in 1961 and restated in 1974 by the Federal Council of Science and Technology, states: "The publication of research results is an essential part of the research process. This has been recognized in part through authorization to pay publication costs from federal research grant and contract funds."

It is not necessary for the plumber to write about pipes, nor is it necessary for the lawyer to write about cases (except *brief* writing), but the research scientist, perhaps uniquely among the trades and professions, must provide a written document showing what he or she did, why it was done, how it was done, and what was learned from it.

Thus the scientist must not only "do" science but must "write" science. Although good writing does not lead to the publication of bad science, bad writing can and often does prevent or delay the publication of good science. Unfortunately, the education and training of scientists are often so overwhelmingly committed to science that the communication arts are neglected or ignored. In short, many good scientists are poor writers. Certainly, many scientists do not like to write. As Charles Darwin said, "a naturalist's life would be a happy one if he had only to observe and never to write" (quoted by Trelease [38]).

The purpose of this book is to help scientists and students of the sciences in all disciplines to prepare manuscripts that will have a high probability of being accepted for publication and of being completely understood when they are published. Because the requirements of journals vary widely from discipline to discipline, and even within the same discipline, it is not possible to offer recommendations that are universally acceptable. In this book, I attempt, however, to present certain basic principles that are accepted in most disciplines.

The genesis of this book occurred some 25 years ago when I taught a graduate seminar in scientific writing at the Institute of Microbiology at Rutgers University. I quickly learned that graduate students in the sciences both wanted and needed *practical* information about writing. If I lectured about the pros and cons of split infinitives, my students became somnolent; if I lectured about how to organize data into a table, they were wide awake. For that reason, I used a straightforward "how to" approach when I later published an article (16) based on my old lecture notes. The article turned out to be surprisingly popular and that led naturally to the publication of the First Edition of this book.

And the First Edition led naturally to this Second Edition. Because this book is now being used in teaching programs in several hundred colleges and universities, it seems desirable to keep it up to date and to correct the (few) imperfections that appeared in the First Edition. I thank those readers who kindly provided me with comments and criticisms of the First Edition, and I herewith invite additional suggestions and comments that may improve future editions of this book.

Although this Second Edition is somewhat larger and better (he says) than the First, the basic outline of the book has not been altered. Because the reviews of the First Edition were almost universally favorable, drastic revision seemed unwise. And the reviews *were* favorable. One reviewer described the book as "both good and original." Unfortunately, he went on to add that "the part that is good is not original and the part that is original is not good." Several other reviewers compared my writing style with that of Shakespeare, Dickens, and Thackeray—but not favorably.

Without meaning to knock the competition, I should observe that my book is clearly a "how to" book, whereas most other books on the subject of scientific writing are written in more general terms, with emphasis on the language of science. This book was written from the perspective of my many years of experience as a managing editor and as a publisher. Thus, the contents are specific and practical.

In writing this book, I had four goals in mind. First, I delayed writing and publishing it until I was reasonably sure that I would not violate the managing editors' creed: "Don't start vast projects with half-vast ideas." Second, I wanted to present certain information about the scientific paper itself and how to cook it. (Yes, this *is* a cookbook). Third, although this book is in no sense a substitute for a course in English grammar, I do comment repeatedly on the use and misuse of English, with such comments interspersed throughout each chapter and with a summary view of the subject in a later chapter. Fourth, because books such as this are usually as dull as dust, dull to read and dull to write, I have also tried to make the reader laugh. Scientific writing abounds with egregious bloopers (what the British sometimes call "bloomers"), and through the years I have amassed quite a collection of these scientific and grammatical monstrosities, which I am now pleased to share. I have tried to enjoy writing this book, and I hope that you will enjoy reading it.

Note that I say "reading it," even though earlier I described this book as a cookbook. If it were simply a book of recipes, it would hardly be suitable for cover-to-cover reading. Actually, I have tried to organize this material in such a way that it reads logically from start to finish, while at the same time it provides the recipes needed to cook the scientific paper. I hope that users of this book might at least consider a straightforward reading of it. In this way, the reader, particularly the graduate student and fledgling writer, may get something of the flavor of just what a scientific paper is. Then, the book can be used basically as a reference whenever questions arise. The book has a detailed subject index, which should lead the user directly to the appropriate point in the text, whether the needed information is philosophical (e.g., ethical issues involved in dual publication) or technical (e.g., how to crop a photograph for printing).

In the first chapter, I attempt to define a scientific paper. To write a scientific paper, the writer *must* know exactly *what* to do and *why*. Not only does this make the job manageable, but this is precisely the knowledge that the practicing scientist must have, and always keep in mind, to avoid the pitfalls that have ruined the reputations of many scientist authors. To be guilty of dual publication, or to use the work of others without appropriate attribution, is the type of breach in scientific ethics that is regarded as unforgivable by one's peers. Therefore, exact definition of what may go into a scientific paper, and what may not, is of prime importance.

In the next 10 chapters, each individual element of the scientific paper is analyzed, item by item. A scientific paper is the sum of its parts, so let us take a detailed look at each of the component parts. Fortunately, for student and practicing scientist alike, there are certain commonly accepted rules regarding the construction of the title, the Abstract, the Introduction, etc., so that this parts analysis, once mastered, should serve the scientist throughout his or her research career.

In later chapters, associated information is given. Some of this information is technical (how to prepare illustrative material, for example) and some of it is related to the post-writing stages (submission, review, and publishing processes). Then, briefly, the rules relating to primary scientific papers are adjusted to fit different circumstances, such as the writing of review papers, conference reports, book reviews, and theses. Finally, in the last four chapters, I try to present some of the rules of English as applied to scientific writing, a sermon against jargon, a discussion of abbreviations, and a sermon against sin.

At the back of the book are the appendixes, the References, and the Index. As to the references, note that I have used two forms of citation in this book. When I cite something of only passing interest—e.g., a defective title of a published article—the citation is given briefly and parenthetically in the text. Articles and books containing substantial information on the subject under discussion are cited by number only in the text, and the full citations are given in the References at the back of the book. Serious students may wish to consult some of these References for additional or related information.

I do not have all the answers. I thought I did when I was a bit younger. Perhaps I can trace some of my character development to the time when Dr. Smith submitted to one of my journals a surprisingly well-written, well-prepared manuscript, his previous manuscripts having been poorly written, badly organized messes. After review of the new manuscript, I wrote "Dr. Smith, we are happy to accept your superbly written paper for publication in the *Journal*." However, I just couldn't help adding: "Tell me, who wrote it for you?"

Dr. Smith answered: "I am so happy that you found my paper acceptable, but tell me, who read it to you?"

Thus, with appropriate humility, I will try to tell you a few things that may be of use in writing scientific papers.

In the Preface to the First Edition, I stated that I would "view the book as a success if it provides you with the information needed to write effective scientific papers and if it makes me rich and famous." In the first draft of this Preface I proudly wrote that all of these goals had been met. However, upon reading that draft, my wife told me "You had better not say that. If they look you up in Bradstreet, you will be Dun."

Acknowledgments

In most of mankind gratitude is merely
a secret hope for greater favours.
—Duc de la Rochefoucauld

Like a cookbook, a "how to" book presents many recipes that the author has collected over the years. A few of the recipes may be original. Some may be variations of someone else's originals. Many of the recipes in such a collection, however, are "borrowed" intact from other sources.

In this book, I have done a reasonable job, I think, in citing the sources of material borrowed from the published literature. But how about the many ideas and procedures that one has picked up from discussions with colleagues? After the passage of time, one can no longer remember who originated what idea. After the passage of even more time, it seems to me that all of the really good ideas originated with me, a proposition that I know is indefensible.

I am indebted to my friends and colleagues who served with me on the Publications Board of the American Society for Microbiology during the 19 years I served that Society. I am also grateful to the Society for Scholarly Publishing and the Council of Biology Editors, the two organizations from which I have learned the most about scientific writing and publishing.

I am grateful to a number of colleagues who have read the manuscript for this Second Edition and offered valuable comments: Robert E. Bjork, Estella Bradley, L. Leon Campbell, Barton D. Day, Robin A. Day, Jay L. Halio, Edward J. Huth, Karen Kietzman, Evelyn S. Myers, Allie C. Peed, Jr., and Nancy Sakaduski. I am especially grateful to Betty J. Day, without whom this book and my life would make little sense.

Chapter 1

What Is a Scientific Paper?

I'm very good at integral and differential calculus,
I know the scientific names of beings animalculous;
In short, in matters vegetable, animal, and mineral,
I am the very model of a modern Major-General.
                              —W. S. GILBERT

Definition of a Scientific Paper

A scientific paper is a written and published report describing original research results. That short definition must be qualified, however, by noting that a scientific paper must be written in a certain way and it must be published in a certain way, as defined by three centuries of developing tradition, editorial practice, scientific ethics, and the interplay of printing and publishing procedures.

To properly define "scientific paper," we must define the mechanism that creates a scientific paper, namely, valid publication. Abstracts, theses, conference reports, and many other types of literature are published, but such publications do not normally meet the test of valid publication. Further, even if a scientific paper meets all of the other tests, it is not validly published if it is published in the wrong place. That is, a relatively poor research report, but one that meets the tests, is validly published if accepted and published in the right place (a primary journal, usually); a superbly prepared research report is not validly published if published in the wrong place. Most of the government report literature and conference literature, as well as house organs and other ephemeral publications, do not qualify as primary literature.

1

Many people have struggled with the definition of "valid publication," from which is derived the definition of "scientific paper." The Council of Biology Editors (CBE), an authoritative professional organization (in biology, at least) dealing with such problems, arrived at the following definition (13):

> An acceptable primary scientific publication must be the first disclosure containing sufficient information to enable peers (1) to assess observations, (2) to repeat experiments, and (3) to evaluate intellectual processes; moreover, it must be susceptible to sensory perception, essentially permanent, available to the scientific community without restriction, and available for regular screening by one or more of the major recognized secondary services (e.g., currently, Biological Abstracts, Chemical Abstracts, Index Medicus, Excerpta Medica, Bibliography of Agriculture, etc., in the United States and similar facilities in other countries).

At first reading, this definition may seem excessively complex, or at least verbose. But those of us who had a hand in drafting it weighed each word carefully, and we doubt that an acceptable definition could be provided in appreciably fewer words. Because it is important that students, authors, editors, and all others concerned understand what a scientific paper is and what it is not, let us work our way through this definition to see what it really means.

"An acceptable primary scientific publication" starts out as the defined substantive, but this gives way to "the first disclosure," which the rest of the paragraph defines. Certainly, first disclosure of new research data often takes place via oral presentation at a scientific meeting. But the thrust of the CBE statement is that disclosure is more than disgorgement by the author; effective first disclosure is accomplished *only* when the disclosure takes a form that allows the peers of the author (either now or in the future) to comprehend that which is disclosed.

Thus, sufficient information must be presented so that potential users of the data can (i) assess observations, (ii) repeat experiments, and (iii) evaluate intellectual processes. (Are the author's conclusions justified by the data?) Then, the disclosure must be "susceptible to sensory perception." This may seem an awkward phrase, because in normal practice it simply means publication; however, this definition provides for disclosure not just in terms of visual materials (printed journals, microfilm, microfiche) but also perhaps in nonprint, nonvisual forms. For example, "publication" in the form of audio cassettes, if that publication met the other tests provided in the definition, would constitute effective publication. In the future, it is quite possible that first disclosure will be entry into a computer data base.

Regardless of the form of publication, that form must be essentially permanent, must be made available to the scientific community without restriction, and must be made available to the information retrieval services (*Biological Abstracts, Chemical Abstracts, Index Medicus, Science Citation Index,* etc.). Thus, publications such as newsletters and house organs, many of which are of value for their news or other features, cannot serve as repositories for scientific knowledge.

To restate the CBE definition in simpler but not more accurate terms, a scientific paper is (i) the first publication of original research results, (ii) in a form whereby peers of the author can repeat the experiments and test the conclusions, and (iii) in a journal or other source document readily available within the scientific community. To understand this definition, however, we must add an important caveat. In modern science (since about the 1930s), the part of the definition that refers to "peers of the author" is accepted as meaning prepublication peer review. Thus, by definition, scientific papers are published in peer-reviewed publications.

I have belabored this question of definition for two very good reasons. First, the entire community of science has long labored with an inefficient, costly system of scientific communication precisely because it (authors, editors, publishers) has been unable or unwilling to define primary publication. As a result, much of the literature is buried in meeting abstracts, obscure conference reports, government documents, or in books or journals of minuscule circulation. Other papers, in the same or slightly altered form, are published twice or more; occasionally, this is the result of poor ethics on the part of the author, but more often it is the lack of definition as to which conference reports, books, and compilations are (or should be) primary publications and which are not. Redundancy and confusion result.

Second, a scientific paper is, by definition, a particular kind of document containing certain specified kinds of information. A scientific paper "demands exactly the same qualities of thought as are needed for the rest of science: logic, clarity, and precision" (40). If the graduate student or the budding scientist (and even some of those scientists who have already published many papers) can fully grasp the significance of this definition, the writing task should be a good deal easier. Confusion results from an amorphous task. The easy task is the one in which you know exactly what must be done and in exactly what order it must be done.

Organization of a Scientific Paper

A scientific paper is a paper organized to meet the needs of valid publication. A scientific paper is, or should be, highly stylized, with distinctive and clearly evident component parts. Each scientific paper should have, in proper order, its Introduction, Materials and Methods, Results, and Discussion. Any other order will pose hurdles for the reader and probably the writer. "Good organization is the key to good writing" (29).

I have taught and recommended this prescribed order for many years. Until recently, however, there have been several somewhat different systems of organization that were preferred by some journals and some editors. The tendency toward uniformity has increased since 1972, when the order cited above was prescribed as a standard (4).

This order is so eminently logical that, increasingly, it is used for many other types of expository writing. Whether one is writing an article about chemistry, archeology, economics, or crime in the streets, an effective way to proceed is to answer these four questions, in order: (i) What was the problem? Your answer is the *Introduction*. (ii) How did you study the problem? Your answer is the *Materials and Methods*. (iii) What did you find? Your answer is the *Results*. (iv) What do these findings mean? Your answer is the *Discussion*.

What I have said above is generally true for papers reporting laboratory studies. There are, of course, exceptions. As examples, reports of field studies in the earth sciences and clinical case reports in the medical sciences do not readily lend themselves to this kind of organization. However, even in these "descriptive" papers, the same logical progression from problem to solution is often appropriate.

Occasionally, the organization of even "laboratory" papers must be different. If a number of methods were used to achieve directly related results, it might be desirable to combine the Materials and Methods and the Results into an integrated "Experimental" section. Rarely, the results might be so complex or provide such contrasts that immediate discussion seems necessary, and a combined Results and Discussion section might then be desirable.

In descriptive areas of science, there is a wide variety of types of organization. To determine how to organize such papers, and which general headings to use, you will need to refer to the Instructions to Authors of your target journal. If you are in doubt as to the journal, or if the journal publishes widely different kinds of papers, you can obtain general information from appropriate source books. For example, the several major types of medical papers are described in detail by Huth

(21), and the many types of engineering papers and reports are outlined by Michaelson (25).

The well-written scientific paper should report its original data in an organized fashion and in appropriate language. In the chapters that follow, I examine each aspect of that organization, while commenting on language along the way.

In short, I take the position that the preparation of a scientific paper has almost nothing to do with literary skill. It is a question of *organization*. A scientific paper is not "literature." The preparer of a scientific paper is not really an "author" in the literary sense. In fact, I go so far as to say that, if the ingredients are properly organized, the paper will virtually write itself.

Some of my old-fashioned colleagues think that scientific papers should be literature, that the style and flair of an author should be clearly evident, and that variations in style encourage the interest of the reader. I disagree. I think scientists should indeed be interested in reading literature, and perhaps even in writing literature, but the communication of research results is a more prosaic procedure. As Booth (9) put it, "Grandiloquence has no place in scientific writing."

Today, the average scientist, to keep up with a field, must examine the data reported in a very large number of papers. Therefore, scientists and, of course, editors must demand a system of reporting data that is uniform, concise, and readily understandable.

I once heard it said: "A scientific paper is not designed to be read. It is designed to be published." Although this was said in jest, there is much truth to it. And, actually, if the paper is designed to be published, it will also be in a prescribed form that can be read and its contents grasped quickly and easily by the reader.

Language of a Scientific Paper

In addition to organization, the second principal ingredient of a scientific paper should be appropriate language. In this book, I keep emphasizing proper use of English, because most scientists have trouble in this area.

If scientific knowledge is at least as important as any other knowledge, then it must be communicated effectively, clearly, in words of certain meaning. The scientist, to succeed in this endeavor, must therefore be literate. David B. Truman, when he was Dean of Columbia College, said it well: "In the complexities of contemporary existence the specialist who is trained but uneducated, technically skilled but culturally incompetent, is a menace."

Although the ultimate goal of scientific research is publication, it has always been amazing to me that so many scientists neglect the responsibilities involved. A scientist will spend months or years of hard work to secure data, and then unconcernedly let much of their value be lost because of lack of interest in the communication process. The same scientist who will overcome tremendous obstacles to carry out a measurement to the fourth decimal place will be in deep slumber while a secretary is casually changing micrograms per milliliter to milligrams per milliliter and while the printer slips in an occasional pounds per barrel.

English need not be difficult. In scientific writing, we say: "The best English is that which gives the sense in the fewest short words" (a dictum printed for some years in the "Instructions to Authors" of the *Journal of Bacteriology*). Literary tricks, metaphors and the like, divert attention from the message to the style. They should be used rarely, if at all, in scientific writing. Justin Leonard, assistant conservation director of Michigan, once said: "The Ph.D. in science can make journal editors quite happy with plain, unadorned, eighth-grade level composition" (*BioScience*, September 1966).

Other Definitions

Now we have defined a scientific paper, which was the intent of this chapter. Before we proceed, however, a few related definitions may be helpful.

Let us preserve "scientific paper" as the term for an original research report. How should this be distinguished from research reports that are not original, or not scientific, or somehow fail to qualify as scientific papers? Let us look at several specific terms: "review paper," "conference report," and "meeting abstract."

A review paper may review almost anything, most typically the recent work in a defined subject area or the work of a particular individual or group. Thus, the review paper is designed to summarize, analyze, evaluate, or synthesize information that *has already been published* (research reports in primary journals). Although much or all of the material in a review paper has previously been published, the spectre of dual publication does not normally arise because the review nature of the work is usually obvious (often in the title of the publication, such as *Microbiological Reviews, Annual Review of Biochemistry*, etc.).

A conference report is a paper published in a book or journal as part of the proceedings of a symposium, national or international congress, workshop, round table, or the like. Such conferences are normally not designed for the presentation of original data, and the resultant proceedings (in a book or journal) do not qualify as primary publications. Conference presentations are often review papers, presenting reviews of the recent work of particular scientists or recent work in particular laboratories. Some of the material reported at some conferences (especially the exciting ones) is in the form of preliminary reports, in which new, original data are reported, often accompanied by interesting speculation. But, usually, these preliminary reports do not qualify, nor are they intended to qualify, as scientific papers. Later, often much later, such work is validly published in a primary journal; by this time, the loose ends have been tied down, all essential experimental details are recorded (so that a competent worker could repeat the experiments), and the speculations are now recorded as conclusions.

Therefore, the vast conference literature that appears in print normally is not *primary*. If original data are presented in such contributions, the data can and should be published (or republished) in an archival (primary) journal. Otherwise, the information may be in essence lost. If publication in a primary journal follows publication in a conference report, there may be copyright and permission problems affecting portions of the work (*see* Chapter 25), but the more fundamental problem of dual publication normally does not and should not arise.

Meeting abstracts, like conference proceedings, are of several widely varying types. Conceptually, however, they are similar to conference reports in that they can and often do contain original information. They are not primary publications, nor should publication of an abstract be considered a bar to later publication of the full report.

In the past, there has been little confusion regarding the typical one-paragraph abstracts published as part of the program or distributed along with the program of a national meeting or international congress. It was usually understood that the papers presented at these meetings would later be submitted for publication in primary journals. More recently, however, there has been a strong trend towards extended abstracts (or "synoptics"). Because publishing all of the full papers presented at a large meeting, such as a major international congress, is very expensive, and because such publication is still not a substitute for the valid publication offered by the primary journal, the movement to extended abstracts makes a great deal of sense. The extended abstract can supply virtually as much information as a full paper; basically, what it lacks is the experimental detail. However, precisely because it lacks experimental detail, it cannot qualify as a scientific paper.

Those of us who are involved with publishing these materials see the importance of careful definition of the different types of papers. More and more publishers, conference organizers, and individual scientists are beginning to agree on these basic definitions and their general acceptance will greatly clarify both primary and secondary communication of scientific information.

Chapter 2

How to Prepare the Title

Of all my verse, like not a line;
But like my title, for it is not mine.
That title from a better man I stole;
Ah, how much better, had I stol'n the whole!
—Robert Louis Stevenson

Importance of the Title

In preparing a title for a paper, the author would do well to remember one salient fact: That title will be read by thousands of people. Perhaps few people, if any, will read the entire paper, but many people will read the title, either in the original journal or in one of the secondary (abstracting and indexing) services. Therefore, all words in the title should be chosen with great care, and their association with one another must be carefully managed. Perhaps the most common error in defective titles, and certainly the most damaging in terms of comprehension, is faulty syntax (word order).

What is a good title? I define it as the fewest possible words that adequately describe the contents of the paper.

Remember that the indexing and abstracting services depend heavily on the accuracy of the title. An improperly titled paper may be virtually lost and never reach its intended audience.

Length of the Title

In my experience, a *few* titles are too short. A paper was submitted to the *Journal of Bacteriology* with the title "Studies on *Brucella*." Obviously, such a title was not very helpful to the potential reader. Was the study taxonomic, genetic, biochemical, or medical? We would certainly want to know at least that much.

Many titles are too long. An overly long title is often less meaningful than a short title. A generation or so ago, when science was less specialized, titles tended to be long and nonspecific, such as "On the addition to the method of microscopic research by a new way of producing colour-contrast between an object and its background or between definite parts of the object itself" (J. Rheinberg, J. R. Microsc. Soc. *1896*:373). That certainly sounds like a poor title; perhaps it would make a good abstract. Overly long titles remind me of a time, back in the days when I was a librarian, when two students were examining the latest additions to the current-journal shelf. One said to the other: "Say, did you read this paper in *Nature* on ribosome structure?" The other student said: "Yes, I read the paper, but I haven't finished the title yet."

Without question, most overly long titles are overly long for only one reason, and it is a very poor one: the use of "waste" words. Often, these waste words appear right at the start of the title, words such as "Studies on," "Investigations on," and "Observations on."

Need for Specific Titles

Let us analyze a sample title: "Action of Antibiotics on Bacteria." Is it a good title? In *form* it is; it is short and carries no excess baggage (waste words). Certainly, it would not be improved by changing it to "Preliminary Observations on the Effect of Certain Antibiotics on Various Species of Bacteria." However, and this brings me to my next point, most titles that are too short are too short for only one reason: the use of general rather than specific terms.

Obviously, we can safely assume that the study introduced by the above title did *not* test the effect of *all* antibiotics on *all* kinds of bacteria. Therefore, the title is essentially meaningless. If only one or a few antibiotics were studied, they should be individually listed in the title. If only one or a few organisms were tested, they should be individually listed in the title. If the number of antibiotics or organisms was awkwardly large for listing in the title, perhaps a group name could have been substituted. Examples of more acceptable titles than our sample are:

"Action of Streptomycin on *Mycobacterium tuberculosis*"
"Action of Streptomycin, Neomycin, and Tetracycline on Gram-Positive Bacteria"

"Action of Polyene Antibiotics on Plant-Pathogenic Bacteria"
"Action of Various Antifungal Antibiotics on *Candida albicans*
and *Aspergillus fumigatus*"

Although these titles are more acceptable than the sample, they are
not especially good because they are still too general. If the "Action of"
can be defined easily, the meaning might be clearer. For example, the
first title above might be phrased "Inhibition of Growth of *Mycobacterium tuberculosis* by Streptomycin."

Long ago, Leeuwenhoek used the word "animalcules," a descriptive
but not very specific word. In the 1930s, Howard Raistrick published
an important series of papers under the title "Studies on Bacteria." A
similar paper today would have a much more specific title. If the study
featured an organism, the title would give the genus and species and
possibly even the strain number. If the study featured an enzyme of an
organism, the title would not be anything like "Enzymes in Bacteria."
It should be something like "Dihydrofolate Reductase in *Bacillus subtilis.*"

Importance of Syntax

In titles, be especially careful of syntax. Most of the grammatical
errors in titles are due to faulty word order.

A paper was submitted to the *Journal of Bacteriology* with the title
"Mechanism of Suppression of Nontransmissible Pneumonia in Mice
Induced by Newcastle Disease Virus." Unless this author had somehow
managed to demonstrate spontaneous generation, it must have been the
pneumonia that was induced and not the mice. (The title should have
read: "Mechanism of Suppression of Nontransmissible Pneumonia Induced in Mice by Newcastle Disease Virus.")

If you no longer believe that babies result from a visit by the stork, I
offer this title (Bacteriol. Proc., p. 102, 1968): "Multiple Infections
Among Newborns Resulting from Implantation with *Staphylococcus
aureus* 502A." (Is this the "Staph of Life"?) I don't mean to worry
female readers of this book, but there seem to be yet other ways to get
pregnant. An article in *Science* (*190*:865, 1977) on the subject of
marijuana had this to say: "But even most of the firmest advocates of
change in the marijuana laws agree that every possible effort should be
made to discourage teenagers and young women who are or may be
pregnant from using the drug."

Another example I stumbled on one day (Clin. Res. *8*:134, 1960):
"Preliminary Canine and Clinical Evaluation of a New Antitumor

Agent, Streptovitacin." When that dog gets through evaluating streptovitacin, I've got some work I'd like to have him look over.

As a grammatical aside, I would encourage you to be careful when you use "using." The word "using" is, I believe, the most common dangling participle in scientific writing. Either there are some more smart dogs, or "using" is misused in this sentence from a recent manuscript: "Using a fiberoptic bronchoscope, dogs were immunized with sheep red blood cells."

Dogs aren't the only smart animals. A manuscript was submitted to the *Journal of Bacteriology* under the title "Isolation of Antigens from Monkeys Using Complement-Fixation Techniques."

Even bacteria are smart. A manuscript was submitted to the *Journal of Clinical Microbiology* under the title "Characterization of Bacteria Causing Mastitis by Gas-Liquid Chromatography." Isn't it wonderful that bacteria can use GLC?

The Title as a Label

The title of a paper is a "label." It is not a sentence. Because it is not a sentence, with the usual subject, verb, object arrangement, it is really simpler than a sentence (or, at least, usually shorter), but the order of the words becomes even more important.

Actually, a few journals do permit a title to be a sentence. Here is an example: "β-Endorphin Is Associated with Overeating in Genetically Obese Mice (*ob/ob*) and Rats (*fa/fa*)" (Science *202*:988, 1978). I suppose this is only a matter of opinion, but I would object to such a title on two grounds. First, the verb ("Is") is a waste word, in that it can be readily deleted without affecting comprehension. Second, inclusion of the "is" results in a title that now seems to be a loud assertion. It has a dogmatic ring to it because we are not used to seeing authors present their results in the present tense, for reasons that are fully developed in Chapter 26.

The meaning and order of the words in the title are of importance to the potential reader who sees the title in the journal table of contents. But these considerations are equally important to *all* potential users of the literature, including those (probably a majority) who become aware of the paper via secondary sources. Thus, the title should be useful as a label accompanying the paper itself, and it also should be in a form suitable for the machine-indexing systems used by *Chemical Abstracts, Science Citation Index, Index Medicus*, and others. Most of the indexing and abstracting services are geared to "key word" systems, generating either KWIC (key word in context) or KWOC (key word out of context) entries. Therefore, it is fundamentally important that the author provide the right "keys" to the paper when labeling it. That is, the terms in the title should be limited to those words that highlight the significant content of the paper in terms that are both understandable and retrievable.

As an aid to readers, "running titles" or "running heads" are printed at the top of each page. Often, the title of the journal or book is given at the top of left-facing pages and the article or chapter titles are given at the top of right-facing pages (as in this book). Usually, a short version of the title is needed because of space limitations. (The maximum character count is likely to be given in the journal's Instructions to Authors.) It is wise to suggest an appropriate running title (on the title page of the manuscript).

Abbreviations and Jargon

Titles should almost never contain abbreviations, chemical formulas, proprietary (rather than generic) names, jargon, and the like. In designing the title, the author should ask: "How would I look for this kind of information in an index?" If the paper concerns an effect of hydrochloric acid, should the title include the words "hydrochloric acid" or should it contain the much shorter and readily recognizable "HCl"? I think the answer is obvious. Most of us would look under "hy" in an index, not under "hc." Furthermore, if some authors used (and journal editors permitted) HCl and others used hydrochloric acid, the user of the bibliographic services might locate only part of the published literature, not noting that additional references are listed under another, abbreviated, entry. Actually, the larger secondary services have computer programs which are capable of bringing together entries such as deoxyribonucleic acid, DNA, and even ADN (*acide deoxyribonucleique*). However, by far the best rule for authors (and editors) is to avoid abbreviations in titles. And the same rule should apply to proprietary names, jargon, and unusual or outdated terminology,

Series Titles

Most editors I have talked to are also opposed to main title-subtitle arrangements and to hanging titles. The main title-subtitle (series) arrangement was quite common some years ago. (Example: "Studies on Bacteria. IV. Cell Wall of *Staphylococcus aureus*.") Today, many editors believe that it is important, especially for the reader, that each published paper "should present the results of an independent, cohesive study; numbered series papers are discouraged" ("Instructions to Authors," *Journal of Bacteriology*). Series papers, in the past, have had a tendency to relate to each other too closely, giving only bits and pieces with each contribution; thus, the reader was severely handicapped unless the whole series could be read consecutively. Furthermore, the series system is annoying to editors because of scheduling problems and delays. (What happens when no. IV is accepted but no. III is rejected or hung up in review?) Additional objections are that a series title almost always provides considerable redundancy; the first part (before the roman numeral) is usually so general as to be useless; and the results when the secondary services spin out a KWIC index are often unintelligible, it being impossible to reconstruct such double titles. (Article titles phrased as questions also become unintelligible, and in my view "question" titles should not be used.)

The hanging title (same as a series title except that a colon substitutes for the roman numeral) is considerably better, avoiding some of the problems mentioned above, but certainly not the peculiar results from KWIC indexing. Unfortunately, a leading scientific journal, *Science*, is a proponent of hanging titles, presumably on the grounds that it is important to get the most important words of the title up to the front. (Example: "Natural and Synthetic Allatotoxins: Suicide Substrates for Juvenile Hormone Biosynthesis"—Science *217*:647, 1982.) Occasionally, hanging titles may be an aid to the reader, but in my opinion they appear pedantic, often place the emphasis on a general term rather than a more significant term, necessitate punctuation, scramble indexes, and in general provide poorer titles than a straightforward label title.

Use of a straightforward title does not lessen the need for proper syntax, however, or for the proper form of each word in the title. For example, a title reading "New Color Standard for Biology" would seem to indicate the development of color specifications for use in describing plant and animal specimens. However, in the title "New Color Standard for Biologists" (BioScience *27*:762, 1977), the new standard might be useful for study of the taxonomy of biologists, permitting us to separate the green biologists from the blue ones.

Chapter 3

How to List the Authors

You don't expect me to know what to say about a play when I don't know who the author is, do you? . . . If it's by a good author, it's a good play, naturally. That stands to reason.

—GEORGE BERNARD SHAW

The Order of the Names

The easiest part of preparing a scientific paper is simply the entering of the bylines: the authors and addresses. Sometimes.

I haven't yet heard of a duel being fought over the order of listing of authors, but I know of instances in which otherwise reasonable, rational colleagues have become bitter enemies solely because they could not agree on whose names should be listed or in what order.

What is the right order? Unfortunately, there are no agreed-upon rules or generally accepted conventions. Some journals (primarily British, I believe) have required that the authors' names be listed in alphabetical order. Such a simple, nonsignificant ordering system has much to recommend it, but the alphabetical system has not yet become common, especially in the United States.

In the past, there has been a general tendency to list the head of the laboratory as an author whether or not he or she actively participated in the research. Often, the "head" was placed last (second of two authors, third of three, etc.). As a result, the terminal spot became a choice placement for its presumed prestige. Thus, two authors, neither of whom was head of a laboratory or even necessarily a senior professor, would vie for the second spot. If there are three or more authors, the "impor-

15

tant" author will want the first or last position, but not in between. Almost all authors want to be listed first on a review paper.

A countervailing and more modern tendency has been to define the *first* author as the senior author and primary progenitor of the work being reported. Even when the first author is a graduate student and the second (third, fourth) author is head of the laboratory, perhaps even a Nobel Laureate, it is now accepted form to refer to the first author as the "senior author" and to assume that he or she did most or all of the research.

The tendency for laboratory directors to insist upon having their own names on all papers published from their laboratories is still with us. So is the tendency to use the "laundry list" approach, naming as an author practically everyone in the laboratory, including technicians who may have cleaned the glassware after the experiments were completed. In addition, the trend toward collaborative research is steadily increasing. Thus, the average number of authors per paper is on the rise.

A new trend, and I think a very good one, is for the established senior scientist to give recognition (first-place listing) to a younger colleague or graduate student. This encouragement of the younger generation probably promotes the continuance of good science, and it is a testament to the character of the senior scientists who graciously engage in the practice.

Definition of Authorship

Perhaps we can now define authorship by saying that the listing of authors should include those, and only those, who actively contributed to the overall design and execution of the experiments. Further, the authors should normally be listed in order of importance *to the experiments*, the first author being acknowledged as the senior author, the second author being the primary associate, the third author possibly being equivalent to the second but more likely having a lesser involvement with the work reported. Colleagues or supervisors should neither ask to have their names on manuscripts nor allow their names to be put on manuscripts reporting research that they themselves have not been intimately involved with. An author of a paper should be defined as one who takes intellectual responsibility for the research results being reported.

Admittedly, resolution of this question is not always easy. It is often incredibly difficult to analyze the intellectual input to a paper. Certainly, those who have worked together intensively for months or years on a research problem might have difficulty in remembering who had the original research concept or whose brilliant idea was the key to the success of the experiments. And what do these colleagues do when everything suddenly falls into place as a result of a searching question by the traditional "guy in the next lab" who had nothing whatever to do with the research?

Each listed author should have made an important contribution to the study being reported, "important" referring to those aspects of the study which produced new information, the concept that defines an original scientific paper.

The sequence of authors on a published paper should be decided, unanimously, before the research is started. A change may be required later, depending on which turn the research takes, but it is foolish to leave this important question of authorship to the very end of the research process.

On occasion, I have seen ten or more authors listed at the head of a paper (sometimes only a Note). For example, a paper by F. Bulos et al. (Phys. Rev. Letters *13:*486, 1964) had 27 authors and only 12 paragraphs. Such papers frequently come from laboratories that are so small that ten people couldn't fit into the lab, let alone make a meaningful contribution to the experiment.

What accounts for the tendency to list a host of authors? There may be several reasons, but the primary one no doubt relates to the publish or perish syndrome. Some workers wheedle or cajole their colleagues so effectively that they become authors of most or all of the papers coming out of their laboratory. Their research productivity might in fact be meager, yet at year's end their publication lists might indeed be extensive. In some institutions, such padded lists might result in promotion. Such padding is also considered good grantsmanship. Nonetheless, the practice is not recommended. Perhaps a few administrators are fooled, and momentary advantages are sometimes gained by these easy riders. But I suspect that *good* scientists do not allow dilution of their own work by adding other people's names for their minuscule contributions, nor do they want their own names sullied by addition of the names of a whole herd of lightweights.

In short, the scientific paper should list as authors only those who contributed *substantially* to the work. The dilution effect of the multi-author approach adversely affects the *real* investigators. (And, as a former Managing Editor, I can't help adding that this reprehensible practice leads to bibliographic nightmares for all of us involved with use and control of the scientific literature.)

Defining the Order: An Example

Perhaps the following example will help clarify the level of conceptual or technical involvement that should define authorship.

Suppose that Scientist A designs a series of experiments that might result in important new knowledge, and then Scientist A instructs Technician B exactly how to perform the experiments. If the experiments work out and a manuscript results, Scientist A should be the sole author, even though Technician B did all the work. (Of course, the assistance of Technician B should be recognized in the Acknowledgments.)

Now let us suppose that the above experiments do not work out. Technician B takes the negative results to Scientist A and says something like, "I think we might get this damned strain to grow if we change the incubation temperature from 24 to 37°C and if we add serum albumin to the medium." Scientist A agrees to a trial, the experiments this time yield the desired outcome, and a paper results. In this case, Scientist A and Technician B, in that order, should both be listed as authors.

Let us take this example one step further. Suppose that the experiments at 37°C and with serum albumin work, but that Scientist A perceives that there is now an obvious loose end; that is, growth under these conditions suggests that the test organism is a pathogen, whereas the previously published literature had indicated that this organism was nonpathogenic. Scientist A now asks colleague Scientist C, an expert in pathogenic microbiology, to test this organism for pathogenicity. Scientist C runs a quick test by injecting the test substance into laboratory mice in a standard procedure that any medical microbiologist would use, and confirms pathogenicity. A few important sentences are then added to the manuscript, and the paper is published. Scientist A and Technician B are listed as authors; the assistance of Scientist C is noted in the Acknowledgments.

Suppose, however, that Scientist C gets interested in this peculiar strain and proceeds to conduct a series of well-planned experiments which lead to the conclusion that this particular strain is not just mouse-pathogenic, but is the long-sought culprit in certain rare human infections. Thus, two new tables of data are added to the manuscript, and the Results and Discussion are rewritten. The paper is then published listing Scientist A, Technician B, and Scientist C as authors.

Proper and Consistent Form

As to names of authors, the preferred designation normally is first name, middle initial, last name. If an author uses only initials, which has been a regrettable tendency in science, the scientific literature may become confused. If there are two people named Jonathan B. Jones, the literature services can probably keep them straight (by addresses). But if dozens of people publish under the name J. B. Jones (especially if, on occasion, some of them use Jonathan B. Jones), the retrieval services have a hopeless task in keeping things neat and tidy.

In general, scientific journals do not print either degrees or titles after authors' names. (You know what "B.S." means. "M.S." is More of the Same. "Ph.D." is Piled Higher and Deeper. "M.D." is Much Deeper.) However, most medical journals do give degrees after the names. Titles are also often listed in medical journals, either after the names and degrees, or in footnotes. Contributors should consult the journal's "Instructions to Authors" or a recent issue regarding preferred usage.

If a journal allows both degrees and titles, perhaps a bit of advertising might be allowed also, as suggested by the redoubtable Leo Rosten (33):

> Dr. Joseph Kipnis—Psychiatrist
> Dr. Eli Lowitz—Proctologist
> Specialists in Odds and Ends.
>
> Dr. M. J. Kornblum and Dr. Albert Steinkoff,
> Obstetricians 24 Hour Service . . . We Deliver.

Chapter 4

How to List the Addresses

Addresses are given to us to conceal our whereabouts.
—"SAKI" (H. H. MUNRO)

The Rules

The rules of listing the addresses are simple but often broken. As a result, authors cannot always be connected with addresses. Most often, however, it is the style of the journal that creates confusion, rather than sins of commission or omission by the author.

With one author, one address is given (the name and address of the laboratory in which the work was done). If, before publication, the author has moved to a different address, the new address should be indicated in a "Present Address" footnote.

When two or more authors are listed, each in a different institution, the addresses should be listed in the same order as the authors.

The primary problem arises when a paper is published by, let us say, three authors from two institutions. In such instances, each author's name and address should include an appropriate designation such as a superior *a*, *b*, or *c* after the author's name and before (or after) the appropriate address.

This convention is often useful to readers who may want to know whether R. Jones is at Yale or at Harvard. Clear identification of authors and addresses is also of prime importance to several of the secondary services. For these services to function properly, they need to know whether a paper published by J. Jones was authored by the J. Jones of Iowa State or the J. Jones of Cornell or the J. Jones of Cambridge

University in England. Only when authors can be properly identified can their publications be grouped together in citation indexes.

Purposes

Remember that an address serves two purposes. It serves to identify the author; it also supplies (or should supply) the author's mailing address. The mailing address is necessary for many reasons, the most common one being to denote the source of reprints. Although it is not necessary as a rule to give street addresses for most institutions, it should be mandatory these days to provide postal codes.

Some journals use a system of asterisks, footnotes, or the Acknowledgments to indicate "the person to whom inquiries regarding the paper should be addressed." Authors should be aware of journal policy in this regard, and they should decide *in advance* who is to purchase and distribute reprints and from what address (since normally it is the institution that purchases the reprints, not the individual).

Unless a scientist wishes to publish anonymously (or as close to it as possible), a full name and a full address should be considered obligatory.

Chapter 5

How to Prepare the Abstract

Good my lord, will you see the players well bestowed? Do you hear, let them be well used; for they are the abstracts and brief chronicles of the time: after your death you were better have a bad epitaph than their ill report while you live.

—WILLIAM SHAKESPEARE

Definition

When I was very young, my daddy told me: "If you can't be an athlete, be an athletic supporter." I was never quite sure what that meant, but perhaps a similar philosophy can be applied to the writing of abstracts: If you can't write a full exposition because of the 250-word limitation, try to uphold the germinal plasm. In other words, forgetting all of the experimental detail, omitting all of the references to previous work (even the critically important papers that you have published), and avoiding all of the lengthy exposition of your detailed knowledge of the general and specific problems investigated, you should limit yourself to a short description of the problem and its solution. As Houghton (20) put it, "An abstract can be defined as a summary of the information in a document."

"A well-prepared abstract enables readers to identify the basic content of a document quickly and accurately, to determine its relevance to their interests, and thus to decide whether they need to read the

22

document in its entirety" (3). The Abstract should not exceed 250 words and should be designed to define clearly what is dealt with in the paper. Many people will read the Abstract, either in the original journal or in *Biological Abstracts, Chemical Abstracts*, or one of the other secondary publications.

The Abstract should (i) state the principal objectives and scope of the investigation, (ii) describe the methodology employed, (iii) summarize the results, and (iv) state the principal conclusions. The importance of the conclusions is indicated by the fact that they should be said three times: once in the Abstract, again in the Introduction, and again (in more detail probably) in the Discussion.

The Abstract should never give any information or conclusion that is not stated in the paper. References to the literature must not be cited in the Abstract (except in rare instances, such as modification of a previously published method).

Types of Abstracts

The above rules apply to the abstracts that are used in primary journals and often without change in the secondary services (*Chemical Abstracts*, etc.). This type of abstract is often referred to as an *informative* abstract, and it is designed to capsulize the paper. It can and should briefly state the problem, the method used to study the problem, and the principal data and conclusions. Often, the abstract supplants the need for reading of the full paper; without such abstracts, scientists would not be able to keep up in active areas of research.

Another common type of abstract is the *indicative* abstract (sometimes called a descriptive abstract). This type of abstract is designed to indicate the content of a paper, essentially serving as a table of contents, making it easy for potential readers to decide whether to read the paper. However, because of its descriptive rather than substantive nature, it can seldom serve as a substitute for the full paper. Thus, indicative abstracts should not be used as "heading" abstracts in research papers, but they may be used in other types of publications (review papers, conference reports, the government report literature, etc.); such indicative abstracts are often of great value to reference librarians.

An effective discussion of the various uses and types of abstracts was provided by McGirr (23), whose conclusions are well worth repeating: "When writing the abstract, remember that it will be published by itself, and should be self-contained. That is, it should contain no bibliographic, figure, or table references ... The language should be familiar to the potential reader. Omit obscure abbreviations and acronyms. Write the paper before you write the abstract, if at all possible."

Economy of Words

Occasionally, a scientist omits something important from the Abstract. By far the most common fault, however, is the inclusion of extraneous detail.

I once heard of a scientist who had some terribly involved theory about the relation of matter to energy. He then wrote a terribly involved paper. However, the scientist, knowing the limitations of editors, realized that the Abstract of his paper would have to be short and simple if the paper were to be judged acceptable. So, he spent hours and hours honing his Abstract. He eliminated word after word until, finally, all of the verbiage had been removed. What he was left with was the shortest Abstract ever written: "$e = mc^2$."

Today, most scientific journals print a heading Abstract with each paper. It is printed (and should be typed) as a single paragraph. Because the Abstract precedes the paper itself, and because the editors and reviewers like a bit of orientation, the Abstract is almost universally the first part of the manuscript read during the review process. Therefore, it is of fundamental importance that the Abstract be written clearly and simply. If you cannot attract the interest of the reviewer in your Abstract, your cause may be lost. Very often, the reviewer may be perilously close to a final judgment of your manuscript after reading the Abstract alone. This could be because the reviewer has a short attention span (often the case). However, if by definition the Abstract is simply a very short version of the whole paper, it is only logical that the reviewer will often reach a premature conclusion, and that conclusion is likely to be the correct one. Usually, a good Abstract is followed by a good paper; a poor Abstract is a harbinger of woes to come.

Because a heading Abstract is required by most journals and because a meeting Abstract is a requirement for participation in a great many national and international meetings (participation sometimes being determined on the basis of submitted abstracts), scientists should master the fundamentals of Abstract preparation. A recent book by Cremmins (14) can be recommended for this purpose.

When writing the Abstract, examine every word with care. If you can tell your story in 100 words, do not use 200. Economically, it doesn't make sense to waste words. It costs about 12 cents a word to publish a scientific paper and another 12 cents every time that word is reprinted in an abstracting publication, and the total communication system can afford only so much verbal abuse. Of more importance to you, the use of clear, significant words will impress the editors and reviewers (not to mention readers), whereas the use of abstruse, verbose constructions is very likely to provoke a check in the "reject" box on the review form.

Or, in Napoleon's last words: "Make mine a short bier."

Chapter 6

How to Write the Introduction

*What we call the beginning is often the end
and to make an end is to make a beginning.
The end is where we start from.*
                                                    —T. S. ELIOT

Suggested Rules

Now that we have the preliminaries out of the way, we come to the
paper itself. I should mention that some experienced writers prepare
their title and Abstract after the paper is written, even though by
placement they come first. You should, however, have in mind (if not
on paper) a provisional title and an outline of the paper that you propose
to "write up." You should also consider the level of the audience you
are writing for, so that you will have a basis for determining which
terms and procedures need definition or description and which do not.
If you do not have clear purposes in mind, you might go writing off in
six directions at once.

It is a wise policy to begin writing the paper while the work is still in
progress. This makes the writing easier because everything is fresh in
mind. Furthermore, the writing process itself is likely to point to
inconsistencies in the results or perhaps to suggest interesting sidelines
that might be followed. Thus, start the writing while the experimental
apparatus and materials are still available. If you have coauthors, it is
wise to write up the work while they are still available for consultation.

The first section of the text proper should, of course, be the Introduction. The purpose of the Introduction should be to supply sufficient background information to allow the reader to understand and evaluate the results of the present study without needing to refer to previous publications on the topic. The Introduction should also provide the rationale for the present study. Above all, you should state briefly and clearly your purpose in writing the paper. Choose references carefully to provide the most important background information.

Suggested rules for a good Introduction are as follows: (i) It should present first, with all possible clarity, the nature and scope of the problem investigated. (ii) It should review the pertinent literature to orient the reader. (iii) It should state the method of the investigation. If deemed necessary, the reasons for the choice of a particular method should be stated. (iv) It should state the principal results of the investigation. Do not keep the reader in suspense; let the reader follow the development of the evidence. An O. Henry surprise ending might make good literature, but it hardly fits the mold that we like to call the scientific method.

Let me expand on that last point. Many authors, especially beginning authors, make the mistake (and it is a mistake) of holding up their most important findings until late in the paper. In extreme cases, authors have sometimes omitted important findings from the Abstract, presumably in the hope of building suspense while proceeding to a well-concealed, dramatic climax. However, this is a sophomoric gambit which, among knowledgeable scientists, goes over like a double negative at a grammarians' picnic. Basically, the problem with the surprise ending is that the readers become bored and stop reading long before they get to the punch line. "Reading a scientific article isn't the same as reading a detective story. We want to know from the start that the butler did it" (31).

Reasons for the Rules

The first three rules for a good Introduction need little expansion, being reasonably well accepted by most scientist-writers, even beginning ones. It is important to keep in mind, however, that the purpose of the Introduction is to *introduce* (the paper). Thus, the first rule (definition of the problem) is the cardinal one. And, obviously, if the problem is not stated in a reasonable, understandable way, readers will have no interest in your solution. Even if the reader labors through your paper, which is unlikely if you haven't presented the problem in a meaningful way, the reader will be unimpressed with the brilliance of your solution. In a sense, a scientific paper is like other types of journalism. In the

Introduction you should have a "hook" to gain the reader's attention. Why did you choose *that* subject, and why is it *important?*

The second and third rules relate to the first. The literature review and choice of method should be presented in such a way that the reader will understand what the problem was and how you attempted to resolve it.

These three rules then lead naturally to the fourth, the statement of principal results, which should be the capstone of the Introduction.

Citations and Abbreviations

If you have previously published a preliminary note or abstract of the work, you should mention this (with the citation) in the Introduction. If closely related papers have been or are about to be published elsewhere, you should say so in the Introduction, customarily at or towards the end. Such references help to keep the literature neat and tidy for those who must search it.

In addition to the above rules, keep in mind that your paper may well be read by people outside your narrow specialty. Therefore, the Introduction is the proper place to define any specialized terms or abbreviations which you intend to use. Let me put this in context by citing a sentence from a letter of complaint I once received. The complaint was in reference to an ad which had appeared in the *Journal of Virology.* The ad announced an opening for a virologist at the National Institutes of Health (NIH), and concluded with the statement "An equal opportunity employer, M & F." The letter suggested that "the designation 'M & F' may mean that the NIH is muscular and fit, musical and flatulent, hermaphroditic, or wants a mature applicant in his fifties."

Chapter 7

How to Write the Materials and Methods Section

Though this be madness, yet there is method in't.
—WILLIAM SHAKESPEARE

Purpose of the Section

In the first section of the paper, the Introduction, you stated (or should have) the methodology employed in the study. If necessary, you also defended the reasons for your choice of a particular method over competing methods.

Now, in Materials and Methods, you must give the full details. The main purpose of the Materials and Methods section is to provide enough detail that a competent worker can repeat the experiments. Many (probably most) readers of your paper will skip this section, because they already know (from the Introduction) the general methods you used and they probably have no interest in the experimental detail. However, careful writing of this section is critically important because the cornerstone of the scientific method *requires* that your results, to be of scientific merit, must be reproducible; and, for the results to be adjudged reproducible, you must provide the basis for repetition of the experiments by others. That experiments are unlikely to be reproduced is beside the point; the potential for producing the same or similar results *must* exist, or your paper does not represent good science.

When your paper is subjected to peer review, a good reviewer will read the Materials and Methods carefully. If there is serious doubt that

29

your experiments could be repeated, the reviewer will recommend rejection of your manuscript no matter how awe-inspiring your results.

Materials

For materials, include the exact technical specifications and quantities and source or method of preparation. Sometimes it is even necessary to list pertinent chemical and physical properties of the reagents used. Avoid the use of trade names; use of generic or chemical names is usually preferred. This avoids the advertising inherent in the trade name. Besides, the nonproprietary name is likely to be known throughout the world, whereas the proprietary name may be known only in the country of origin. However, if there are known differences among proprietary products and if these differences may be critical (as with certain microbiological media), then use of the trade name, plus the name of the manufacturer, is essential.

Experimental animals, plants, and microorganisms should be identified accurately, usually by genus, species, and strain designations. Sources should be listed and special characteristics (age, sex, genetic and physiological status) described. If human subjects are used, the criteria for selection should be described, and an "informed consent" statement should be added to the manuscript if required by the journal.

Because the value of your paper (and your reputation) can be damaged if your results are not reproducible, you must describe research materials with great care. Be sure to examine the Instructions to Authors of the journal to which you plan to submit the manuscript, because there important specifics are often detailed. Below is a carefully worded statement applying to cell lines (taken from the Information for Authors of *In Vitro*, the Journal of the Tissue Culture Association):

> *Cell line data:* The source of cells utilized, species, sex, strain, race, age of donor, whether primary or established, must be clearly indicated. The supplier name, city, and state abbreviation should be stated within parentheses when first cited. Specific tests used for verification of purported origin, donor traits, and detection for the presence of microbial agents should be identified. Specific tests should be performed on cell culture substrates for the presence of mycoplasmal contamination by using both a direct agar culture and an indirect staining or biochemical procedure. A brief description or a proper reference citation of the procedure used must be included. If these tests were not performed, this fact should be clearly stated in the Materials and Methods section. Other data relating to unique biological, biochemical and/or immunological markers should also be included if available.

Methods

For methods, the usual order of presentation is chronological. Obviously, however, related methods should be described together, and straight chronological order cannot always be followed. For example, if a particular assay was not done until late in the research, the assay method should be described along with the other assay methods, not by itself in a later part of Materials and Methods.

Headings

The Materials and Methods section is the first section of the paper in which subheadings should be used. (See Chapter 14 for discussion of the how and when of subheadings.) When possible, construct subheadings that "match" those to be used in Results. The writing of both sections will be easier if you strive for internal consistency, and the reader will be able to grasp quickly the relationship of a particular methodology to the related Results.

Measurements and Analysis

Be precise. If a reaction mixture was heated, give the temperature. Questions such as "how" and "how much" should be precisely answered by the author and not left for the reviewer or the reader to puzzle over.

Statistical analyses are often necessary, but you should feature and discuss the data, not the statistics. Generally, a lengthy description of statistical methods indicates that the writer has recently acquired this information and believes that the readers need similar enlightenment. Ordinary statistical methods should be used without comment; advanced or unusual methods may require a literature citation.

And, again, be careful of your syntax. A recent manuscript described what could be called a disappearing method. The author stated, "The radioactivity in the tRNA region was determined by the trichloroacetic acid-soluble method of Britten et al." And then there is the painful method: "After standing in boiling water for an hour, examine the flask."

Need for References

In describing the methods of the investigations, you should give sufficient details so that a competent worker could repeat the experiments, as stated above. If your method is new (unpublished), you must provide *all* of the needed detail. However, if a method has been previously published in a standard journal, only the literature reference should be given. But I recommend more complete description of the method if the only previous publication was in, let us say, the *South Tasmanian Journal of Nervous Diseases of the Gnat.*

If several alternative methods are commonly employed, it is useful to identify the method briefly as well as to cite the reference. For example, it is preferable to state "cells were broken by ultrasonic treatment as previously described (9)" than to state "cells were broken as previously described (9)."

Tabular Material

When large numbers of microbial strains or mutants are used in a study, prepare strain tables identifying the source and properties of mutants, bacteriophages, plasmids, etc. The properties of a number of chemical compounds can also be presented in tabular form, often to the benefit of both the author and the reader.

A method, strain, etc., used in only one of several experiments reported in the paper, should be described in the Results section or, if brief enough, may be included in a table footnote or a figure legend.

Correct Form and Grammar

Finally, do *not* make the common error of mixing some of the Results in this section.

In summary, there is only one rule for a properly written Materials and Methods section: Enough information must be given so that the experiments could be reproduced by a competent colleague.

A good test, by the way (and a good way to avoid rejection of your manuscript), is to give a copy of your finished manuscript to a colleague and ask if he or she could repeat the experiments. It is quite possible that, in reading about your Materials and Methods, your colleague will pick up a glaring error that you missed simply because you were too close to the work. For example, you might have described your distillation apparatus, procedure, and products with infinite care, and then inadvertently neglected to define the starting material or to state the distillation temperature.

Mistakes in grammar and punctuation are not always serious; the meaning of general concepts, as expressed in the Introduction and Discussion, can often survive a bit of linguistic mayhem. In Materials and Methods, however, exact and specific items are being dealt with and precise use of English is a must. Even a missing comma can cause havoc, as in the sentence: "Employing a straight platinum wire rabbit, sheep and human blood agar plates were inoculated . . ." That sentence was in trouble right from the start, because the first word is a dangling participle. Comprehension didn't totally go out the window, however, until the author neglected to put a comma after "wire."

Because the Materials and Methods section usually gives short, discrete bits of information, the writing sometimes becomes telescopic; details essential to the meaning may then be omitted. The most common error is to state the action without stating the agent of the action. In the sentence "To determine its respiratory quotient, the organism was . . . ," the only stated agent of the action is "the organism," and somehow I doubt that the organism was capable of making such a determination. Here is a similar sentence: "Having completed the study, the bacteria were of no further interest." Again, I doubt that the bacteria "completed the study"; if they did, their lack of "further interest" was certainly an act of ingratitude.

"Blood samples were taken from 48 informed and consenting patients . . . the subjects ranged in age from 6 months to 22 years" (Pediatr. Res. *6*:26, 1972). There is no grammatical problem with that sentence, but the telescopic writing leaves the reader wondering just how the 6-month-old infants gave their informed consent.

And, of course, always watch for spelling errors, both in the manuscript and on the galley proofs. I am not an astronomer, but I suspect that a word is misspelled in the following sentence: "We rely on theatrical calculations to give the lifetime of a star on the main sequence" (Annu. Rev. Astron. Astrophys. *1*:100, 1963).

Chapter 8

How to Write the Results

The great tragedy of Science—
the slaying of a beautiful hypothesis
by an ugly fact.
—T. H. HUXLEY

Content of the Results

So now we come to the core of the paper, the data. This part of the paper is called the Results section.

Contrary to popular belief, you shouldn't start the Results section by describing methods which you inadvertently omitted from the Materials and Methods section.

There are usually two ingredients of the Results section. First, you should give some kind of overall description of the experiments, providing the "big picture," without, however, repeating the experimental details previously provided in Materials and Methods. Second, you should present the data.

Of course, it isn't quite that easy. How do you present the data? A simple transfer of data from laboratory notebook to manuscript will hardly do.

Most important, in the manuscript you should present representative data rather than endlessly *repetitive* data. The fact that you could perform the same experiment 100 times without significant divergence in results might be of considerable interest to your major professor, but editors, not to mention readers, prefer a little bit of predigestion. Woodford (40) stated this concept pungently: "Some scientists seem to

34

believe that the world will be perpetually grateful to them for keeping what amounts to a public diary of their diligent activity, or endlessly titillated by accounts of their violation of each one of Nature's infinite number of maidenheads." Aaronson (1) said it another way: "The compulsion to include everything, leaving nothing out, does not prove that one has unlimited information; it proves that one lacks discrimination." Exactly the same concept, and it is an important one, was stated almost a century earlier by John Wesley Powell, a geologist who served as President of the American Association for the Advancement of Science in 1888. In Powell's words: "The fool collects facts; the wise man selects them."

How to Handle Numbers

If one or only a few determinations are to be presented, they should be treated descriptively in the text. Repetitive determinations should be given in tables or graphs.

Any determinations, repetitive or otherwise, should be meaningful. Suppose that, in a particular group of experiments, a number of variables were tested (one at a time, of course). Those variables which affected the reaction become determinations or data and, if extensive, are tabulated or graphed. Those variables which do not seem to affect the reaction need not be tabulated or presented; however, it is often important to define even the *negative* aspects of your experiments. It is often good insurance to state what you did *not* find under the *conditions of your experiments.* Someone else very likely may find different results under *different conditions.* Carl Sagan (34) said it well: " . . . absence of evidence is not evidence of absence."

If statistics are used to describe the results, they should be meaningful statistics. Erwin Neter, former Editor-in-Chief of *Infection and Immunity*, tells a classic story to emphasize this point. He refers to a paper which reputedly read: "33⅓% of the mice used in this experiment were cured by the test drug; 33⅓% of the test population were unaffected by the drug and remained in a moribund condition; the third mouse got away."

Strive for Clarity

The results should be short and sweet, without verbiage. Mitchell (26) quoted Einstein as having said, "If you are out to describe the truth, leave elegance to the tailor." Although the Results section of a paper is the most important part, it is often the shortest, particularly if it is preceded by a well-written Materials and Methods section and followed by a well-written Discussion.

When the perfect scientific paper is written, if it ever is, the Results section may possibly have just one sentence: "The results are shown in Table 1."

If any part of the paper needs to be clearly and simply stated, it is the Results, because it is the results that comprise the new knowledge that you are contributing to the world. The earlier parts of the paper (Introduction, Materials and Methods) are designed to tell why and how you got the Results; the later part of the paper (Discussion) is designed to tell what they mean. Obviously, therefore, the whole paper must stand or fall on the basis of the Results. Thus, the Results must be presented with crystal clarity.

Avoid Redundancy

Do not be guilty of redundancy in the Results. The most common fault is the repetition in words of what is already apparent to the reader from examination of the figures and tables. Even worse is the actual presentation, in the text, of all or many of the data shown in the tables or figures. This grave sin is committed so frequently that I comment on it at length, with examples, in the chapters on how to prepare the tables and illustrations (Chapters 12 and 13).

Do not be redundant in citing figures and tables. Do not say "It is clearly shown in Table 1 that nocillin inhibited the growth of *N. gonorrhoeae.*" Say "Nocillin inhibited the growth of *N. gonorrhoeae* (Table 1)."

Some writers go too far in avoiding verbiage, however. Such writers often violate the rule of antecedents, the most common violation being the ubiquitous "it." Here is an item from a medical manuscript: "The left leg became numb at times and she walked it off . . . On her second day, the knee was better, and on the third day it had completely disappeared." The antecedent for both "its" was presumably "the numbness," but I rather think that the wording in both instances was a result of dumbness.

Chapter 9

How to Write the Discussion

For undemocratic reasons and for motives not of State,
They arrive at their conclusions—largely inarticulate.
—RUDYARD KIPLING

Discussion and Verbiage

The Discussion is harder to define than the other sections. As a result, it is usually the hardest section to write. And, whether you know it or not, *many* papers are rejected by journal editors because of a faulty Discussion, even though the data of the paper might be both valid and interesting. Even more likely, the true meaning of the data may be completely obscured by the interpretation presented in the Discussion, again resulting in rejection.

Many, if not most, Discussions are too long and verbose. As Doug Savile said, "Occasionally, I recognize what I call the squid technique: the author is doubtful about his facts or his reasoning and retreats behind a protective cloud of ink" (*Tableau*, September 1972).

Some Discussions remind one of the diplomat, described by Allen Drury in *Advise and Consent* (Doubleday & Co., Garden City, NY, 1959, p. 47), who characteristically gave "answers which go winding and winding off through the interstices of the English language until they finally go shimmering away altogether and there is nothing left but utter confusion and a polite smile."

Components of the Discussion

What are the essential features of a good Discussion? I believe the main components will be provided if the following injunctions are

37

heeded:

1. Try to present the principles, relationships, and generalizations shown by the Results. And bear in mind, in a good Discussion, you *discuss—you do not recapitulate* the Results.

2. Point out any exceptions or any lack of correlation, and define unsettled points.

3. Show how your results and interpretations agree (or contrast) with previously published work.

4. Don't be shy; discuss the theoretical implications of your work, as well as any possible practical applications.

5. State your conclusions, as clearly as possible.

6. Summarize your evidence for *each* conclusion. Or, as the wise old scientist will tell you, "Never assume anything except a 6% mortgage."

Factual Relationships

In simple terms, the primary purpose of the Discussion is to show the relationships among observed facts. To emphasize this point, I always tell the old story about the biologist who trained a flea.

After training the flea for many months, the biologist was able to get a response to certain commands. The most gratifying of the experiments was the one in which the professor would shout the command "Jump," and the flea would leap into the air each time the command was given.

The professor was about to submit this remarkable feat to posterity via a scientific journal, but he—in the manner of the true scientist— decided to take his experiments one step further. He sought to determine the location of the receptor organ involved. In one experiment, he removed the legs of the flea, one at a time. The flea obligingly continued to jump upon command, but as each successive leg was removed, its jumps became less spectacular. Finally, with the removal of its last leg, the flea remained motionless. Time after time the command failed to get the usual response.

The professor decided that at last he could publish his findings. He set pen to paper and described in meticulous detail the experiments executed over the preceding months. His conclusion was one intended to startle the scientific world: *When the legs of a flea are removed, the flea can no longer hear.*

Significance of the Paper

Too often, the *significance* of the results is not discussed or not discussed adequately. If the reader of a paper finds himself or herself asking "So what?" after reading the Discussion, the chances are that the author became so engrossed with the trees (the data) that he or she didn't really notice how much sunshine had appeared in the forest.

The Discussion should end with a short summary or conclusion regarding the significance of the work. I like the way Anderson and Thistle (6) said it: "Finally, good writing, like good music, has a fitting climax. Many a paper loses much of its effect because the clear stream of the discussion ends in a swampy delta."

Defining Scientific Truth

In showing the relationships among observed facts, you do not need to reach cosmic conclusions. Seldom will you be able to illuminate the whole truth; more often, the best you can do is shine a spotlight on one area of the truth. Your one area of truth can be buttressed by your data; if you extrapolate to a bigger picture than that shown by your data, you may appear foolish to the point that even your data-supported conclusions are cast into doubt.

One of the more meaningful thoughts in poetry was expressed by Sir Richard Burton in *The Kasidah:*

> All Faith is false, all Faith is true:
> Truth is the shattered mirror strown
> In myriad bits; while each believes
> His little bit the whole to own.

So exhibit your little piece of the mirror, or shine a spotlight on one area of the truth. The "whole truth" is a subject best left to the ignoramuses, who loudly proclaim its discovery every day.

When you describe the meaning of your little bit of truth, do it simply. The simplest statements evoke the most wisdom; verbose language and fancy technical words are used to convey shallow thought.

Chapter 10

How to State the Acknowledgments

Life is not so short but that there is always time enough for courtesy.
—RALPH WALDO EMERSON

Ingredients of the Acknowledgments

The main text of a scientific paper is usually followed by two additional sections, namely, the Acknowledgments and the Literature Cited.

As to the Acknowledgments, two possible ingredients require consideration.

First, you should acknowledge any significant help that you received from any individual, whether in your laboratory or elsewhere. Specifically, you should acknowledge the source of special equipment, cultures, or other materials. Further, you should acknowledge the help of anyone who contributed significantly to the work or to the interpretation of the work.

You might, for example, say something like "Thanks are due to J. Jones for assistance with the experiments and to R. Smith for valuable discussion." Of course, most of us who have been around for awhile recognize that this is simply a thinly veiled way of admitting that Jones did the work and Smith explained what it meant.

Second, it is usually the Acknowledgments wherein you should acknowledge any outside financial assistance, such as grants, contracts, or fellowships. (In these days, you might snidely mention the absence of such grants, contracts, or fellowships.)

40

Being Courteous

The important element in Acknowledgments is simple courtesy. There isn't anything really scientific about this section of a scientific paper. The same rules that would apply in any other area of civilized life should apply here. If you borrowed a neighbor's lawnmower, you would (I hope) remember to say thanks for it. If your neighbor gave you a really good idea for landscaping your property, and you then put that idea into effect, you would (I hope) remember to say thank you. It is the same in science; if your neighbor (your colleague) provided important ideas, important supplies, or important equipment, you should thank him or her. And you must say thanks *in print*, because that is the way that scientific landscaping is presented to its public.

A word of caution is in order. Often, it is wise to show the proposed wording of the Acknowledgment to the person whose help you are acknowledging. He or she might well believe that your acknowledgment is insufficient or (worse) that it is too effusive. If you have been working so closely with an individual that you have borrowed either equipment or ideas, that person is most likely a friend or a valued colleague. It would be silly to risk either your friendship or the opportunities for future collaboration by placing in public print a thoughtless word that might be offensive. An inappropriate thank you can be worse than none at all, and if you value the advice and help of friends and colleagues, you should be careful to thank them in a way that pleases rather than displeases them.

Furthermore, if your acknowledgment relates to an idea, suggestion, or interpretation, be very specific about it. If your colleague's input is too broadly stated, he or she could well be placed in the sensitive and embarrassing position of having to defend the entire paper. Certainly, if your colleague is not a coauthor, you must not make him or her a responsible party to the basic considerations treated in your paper. Indeed, your colleague may not agree with some of your central points, and it is not good science and not good ethics for you to phrase the Acknowledgments in a way that seemingly denotes endorsement.

I wish that the word "wish" would disappear from Acknowledgments. Wish is a perfectly good word when you mean wish, as in "I wish you success." However, the word "wish" is an unacceptable substitute for the word "want." If you say "I wish to thank John Jones," you are wasting words. "I thank John Jones" is sufficient. You may also be introducing the implication that "I wish that I could thank John Jones for his help but it wasn't all that great."

Chapter 11

How to Cite the Literature

In science, read, by preference, the newest works;
in literature, the oldest. The classic literature is
always modern.
—EDWARD BULWER LYTTON

Rules to Follow

There are two rules to follow in the Literature Cited section just as in the Acknowledgments section.

First, you should list only significant, published references. References to unpublished data, papers in press, abstracts, theses, and other secondary materials should not clutter up the Literature Cited. If such a reference seems absolutely essential, you may add it parenthetically or as a footnote in the text.

Second, check all parts of every reference against the original publication, before the manuscript is submitted, and perhaps again at the galley proof stage. Take it from an erstwhile librarian: there are far more mistakes in the Literature Cited section of a paper than anywhere else.

Reference Styles

Journals vary considerably in their style of handling references. One person looked at 52 scientific journals and found 33 different styles for listing references (M. O'Connor, Br. Med. J. *1*(6104):31, 1978). Some journals print titles of articles and some do not. Some insist on inclusive pagination, whereas others print first pages only. The smart author writes out references (on 3 by 5 cards, usually) in full. Then, in preparing a manuscript, he or she has all the needed information. It is easy to edit out information; it is indeed laborious to track down 20 or so references to add article titles or ending pages when required to do so by a journal editor. Even if you know that the journal to which you plan to submit your manuscript uses a short form (no article titles, for example), you would still be wise to establish your reference list in the complete form. This is good practice because (i) the journal you selected may reject your manuscript, resulting in your decision to submit the manuscript to another journal, perhaps one with more demanding requirements, and (ii) it is more than likely that you will use some of the same references again, in later research papers, review articles (and most review journals demand *full* references), or books.

Although there is an almost infinite variety of reference styles, most journals cite references in one of three general ways that are usually referred to as "name and year," "by number from alphabetical list," and "by number in order of citation."

Name and Year System

The name and year system (often referred to as the Harvard system) was very popular for many years and is still used in many journals, although that system is not as widely used as it once was. Its big advantage is convenience to the author. Because the references are unnumbered, references can be added or deleted easily. No matter how many times the reference list is modified, "Smith and Jones (1950)" remains exactly that. If there are two or more "Smith and Jones (1950)" references, the problem is easily handled by listing the first as "Smith and Jones (1950a)," the second as "Smith and Jones (1950b)," etc. The disadvantages of name and year relate to readers and publishers. The disadvantage to the reader occurs when (often in the Introduction) a large number of references must be cited within one sentence or paragraph. Sometimes the reader must jump over several lines of parenthetical references before he or she can again pick up the text. Even two or three references, cited together, can be distracting to the reader. The disadvantage to the publisher is obvious: increased cost. When "Smith, Jones, and Higginbotham (1948)" can be converted to "(7)," considerable composition (typesetting) and printing cost savings can be realized. As already noted, it now costs on the order of 12 cents per word to print scientific journals, so editors (and especially managing editors) are more cost-conscious than ever.

Because some papers have an unwieldy number of authors, most journals that use name and year have an "et al." rule. Most typically, it works as follows. Names are always used in citing papers with either one or two authors, e.g., "Smith (1970)," "Smith and Jones (1970)." If the paper has three authors, list all three the first time the paper is cited, e.g., "Smith, Jones, and McGillicuddy (1970)." If the same paper is cited again, it can be shortened to "Smith et al. (1970)." When a cited paper has four or more authors, it should be cited as "Smith et al. (1970)" even in the first citation. In the Literature Cited section, some journals prefer that all authors be listed (no matter how many); other journals, particularly the leading medical journals, cite only the first three authors and follow with "et al."

Alphabet-Number System

This system, citation by number from an alphabetized list of references, is a modern modification of the name and year system. Citation by numbers keeps printing expenses within bounds; the alphabetized list, particularly if it is a long list, is relatively easy for authors to prepare and readers (especially librarians) to use.

Authors who cut their eye teeth on name and year tend to dislike the alphabet-number system, claiming that citation of numbers cheats the reader. The reader should be told, so the argument goes, the name of the person associated with the cited phenomenon; sometimes, the reader should also be told the date, on the grounds that an 1878 reference might be viewed differently than a 1978 reference.

Fortunately, these arguments can be overcome. As you cite references in the text, decide whether names or dates are important. If they are not (usually the case), use only the reference number: "Pretyrosine is quantitatively converted to phenylalanine under these conditions (13)." If you want to feature the name of the author, do it within the context of the sentence: "The role of the carotid sinus in the regulation of respiration was discovered by Heymans (13)." If you want to feature the date, you can also do that within the sentence: "Streptomycin was first used in the treatment of tuberculosis in 1945 (13)."

Citation Order System

The citation order system is simply a system of citing the references (by number) in the order that they appear in the paper. This system avoids the substantial printing expense of the name and year system, and readers often like it because they can quickly refer to the references if they so desire in one-two-three order as they come to them in the text.

It is a useful system for a journal which is basically a "note" journal, each paper containing only a few references. For long papers, with many references, citation order is probably not a good system. It is not good for the author, because of the substantial renumbering chore that results from addition or deletion of references. It is not ideal for the reader, because the nonalphabetical presentation of the reference list may result in separation of various references to works by the same author.

In the First Edition of this book, I stated that the alphabet-number system "seems to be slowly gaining ascendancy." Soon thereafter, however, the first version of the "Uniform requirements for manuscripts submitted to biomedical journals" appeared, sponsoring the citation order system for the cooperating journals. The "Uniform requirements" (22) have been adopted by several hundred biomedical journals. Thus, it is not now clear which citation system, if any, will gain "ascendancy." The "Uniform Requirements" document is so impressive in so many ways that it has had and is having a powerful impact. In this one area of literature citation, however, there remains strong opposition. For example, the powerful Council of Biology Editors has "decided to use alphabetical listing of the references numbered sequentially in the forthcoming 5th edition of the Style Manual" (CBE Views 5(4):15, 1982). In addition, the long-awaited new edition of *The Chicago Manual of Style* (13th ed., The University of Chicago Press, 1982), the bible of most of the scholarly publishing community, has appeared with its usual ringing endorsement of alphabetically arranged references. In its more than 100 pages of detailed instructions for handling references, it several times makes such comments as (page 431): "The most practical and useful way to arrange entries in a reference list or a bibliography is in alphabetical order, by authors, either running through the whole list or in each section of it."

Titles and Inclusive Pages

Should article titles be given in references? Normally, you will have to follow the style of the journal; if the journal allows a choice (and some do), I recommend that you give *complete* references. By denoting the overall subjects, the article titles make it easy for interested readers (and librarians) to decide whether they need to consult none, some, or all of the cited references.

The use of inclusive pagination (first and last page numbers) makes it easy for potential users to distinguish between one-page notes and 50-page review articles. Obviously, the cost to you (or your library) of obtaining the references, particularly if acquired as photocopies, can vary considerably depending on the number of pages involved.

Journal Abbreviations

Although journal styles vary widely, one aspect of reference citation has been standardized in recent years, i.e., journal abbreviations. As the result of widespread adoption of a standard (2), almost all of the major primary journals and secondary services now use the same system of abbreviation. Previously, most journals abbreviated journal names (significant printing expense can be avoided by abbreviation), but there was no uniformity. The *Journal of the American Chemical Society* was variously abbreviated to "J. Amer. Chem. Soc.," "Jour. Am. Chem. Soc.," "J.A.C.S.," etc. These differing systems posed problems for authors and publishers alike. Now there is essentially only one system, and it is uniform. The word "Journal" is now always abbreviated "J." (Some journals omit the periods after the abbreviations.) By noting a few of the rules, authors can abbreviate many journal titles, even unfamiliar ones, without reference to a source list. It is helpful to know, for example, that all "ology" words are abbreviated at the "l" ("Bacteriology" is abbreviated "Bacteriol."; "Physiology" is abbreviated "Physiol.," etc.). Thus, if one memorizes the abbreviations of words commonly used in titles, most journal titles can be abbreviated with ease. An exception to be remembered is that one-word titles (*Science, Biochemistry*) are never abbreviated. Appendix 1 provides a list of the correct abbreviations for commonly used words in periodical titles.

Citation in the Text

I find it depressing that many authors use slipshod methods in citing the literature. (I never stay depressed long—my attention span is too short.) A common offender is the "handwaving reference," in which the reader is glibly referred to "Smith's elegant contribution" without any hint of what Smith reported or how Smith's results relate to the present author's results. If a reference is worth citing, the reader should be told why.

Some authors get into the habit of putting all citations at the end of sentences. This is wrong. The reference should be placed at that point in the sentence to which it applies. Michaelson (25) gives this good example:

> We have examined a digital method of spread-spectrum modulation for multiple-access satellite communication and for digital mobile radio telephony.[1,2]

Note how much clearer the citations become when the sentence is recast as follows:

> We have examined a digital method of spread-spectrum modulation for use with Smith's development of multiple-access communication[1] and with Brown's technique of digital mobile radiotelephony.[2]

Chapter 12

How to Design
Effective Tables

Now go, write it before them in a
table, and note it in a book.
                    —Isaiah 30:8

When to Use Tables

Before proceeding to the "how to" of tables, let us first examine the question "whether to."

As a general rule, do not construct a table unless repetitive data *must* be presented. There are two reasons for this general rule. First, it is simply not good science to regurgitate reams of data just because you have them in your laboratory notebooks; only samples and breakpoints need be given. Second, the cost of publishing tables is very high (in comparison with straight text), and all of us involved with the generation and publication of scientific literature should worry about the cost.

If you made (or need to present) only a few determinations, give the data in the text. Tables 1 and 2 are useless, yet they are typical of many tables that are submitted to journals.

Table 1 is faulty because two of the columns give standard conditions, not variables and not data. If temperature is a variable in the experiments, it can have its column. If all experiments were done at the same temperature, however, this single bit of information should be noted in Materials and Methods and perhaps as a footnote to the table, but not

TABLE 1. *Effect of aeration on growth of Streptomyces coelicolor*

Temp (°C)	No. of expt	Aeration of growth medium	Growth[a]
24	5	+[b]	78
24	5	−	0

[a] As determined by optical density (Klett units).
[b] Symbols: +, 500-ml Erlenmeyer flasks were aerated by having a graduate student blow into the bottles for 15 min out of each hour; −, identical test conditions, except that the aeration was provided by an elderly professor.

TABLE 2. *Effect of temperature on growth of oak (Quercus) seedlings*[a]

Temp (°C)	Growth in 48 h (mm)
−50	0
−40	0
−30	0
−20	0
−10	0
0	0
10	0
20	7
30	8
40	1
50	0
60	0
70	0
80	0
90	0
100	0

[a] Each individual seedling was maintained in an individual round pot, 10 cm in diameter and 100 m high, in a rich growth medium containing 50% Michigan peat and 50% dried horse manure. Actually, it wasn't "50% Michigan"; the peat was 100% "Michigan," all of it coming from that state. And the manure wasn't half-dried (50%); it was all dried. And, come to think about it, I should have said "50% dried manure (horse)"; I didn't dry the horse at all.

in a column in the table. Thus, Table 1 is not a proper table. These data can be presented in the text itself in a form that is readily comprehensible to the reader, while at the same time avoiding the substantial additional typesetting cost of tabulation. Very simply, these results would read: "Aeration of the growth medium was essential for the growth of *Streptomyces coelicolor*. At room temperature (24°C), no growth was evident in stationary (unaerated) cultures, whereas substantial growth (OD, 78 Klett units) occurred in shaken cultures."

Table 2 has no columns of identical readings, and it looks like a good table. But is it? The independent variable column (temperature) looks reasonable enough, but the dependent variable column (growth) has a suspicious number of zeros. Question any table with a large number of zeros (whatever the unit of measurement) or a large number of 100s when percentages are used. Table 2 is certainly a useless table, because all it tells us is that "The oak seedlings grew at temperatures between 20 and 40°C; no measurable growth occurred at temperatures below 20°C or above 40°C."

In addition to zeros and 100s, be suspicious of plus and minus signs. Table 3 is of a type that often appears in print, although it is obviously not very informative. All this table tells us is that "*S. griseus, S. coelicolor, S. everycolor,* and *S. rainbowenski* grew under aerobic conditions, whereas *S. nocolor* and *S. greenicus* required anaerobic conditions." Whenever a table, or columns within a table, can be readily put into words, do it.

Some authors believe that all data (numbers) must be put in a table. Table 4 is a sad example. It gets even sadder when we learn (at the end of the footnote) that the results were not significant anyway ($P = 0.21$). If these data were worth publishing (which I doubt), one sentence in the Results would have done the job: "The difference between the failure rates—14% (5 of 35) for nocillin and 26% (9 of 34) for potassium penicillin V— was not significant ($P = 0.21$)."

In presenting numbers, give only significant figures. Non-significant figures may mislead the reader by creating a false sense of precision; in addition, comparison of the data becomes more difficult.

Unessential data, such as laboratory numbers, results of simple cal-

TABLE 3. *Oxygen requirement of various species of Streptomyces*

Organism	Growth under aerobic conditions[a]	Growth under anaerobic conditions
Streptomyces griseus	+	−
S. coelicolor	+	−
S. nocolor	−	+
S. everycolor	+	−
S. greenicus	−	+
S. rainbowenski	+	−

[a] See Table 1 for explanation of symbols. In this experiment, the cultures were aerated by a shaking machine (New Brunswick Shaking Co., Scientific, NJ).

TABLE 4. *Bacteriological failure rates*

Nocillin	K Penicillin
5/35 (14)[a]	9/34 (26)

[a] Results expressed as number of failures/total, which is then converted to a percentage (within parentheses). $P = 0.21$.

TABLE 5. *Adverse effects of nicklecillin in 24 adult patients*

No. of patients	Side effect
14	Diarrhea
5	Eosinophilia ($\geqslant 5$ eos/mm^3)
2	Metallic taste[a]
1	Yeast vaginitis[b]
1	Mild rise in urea nitrogen
1	Hematuria (8–10 rbc/hpf)

[a] Both of the patients who tasted metallic worked in a zinc mine.
[b] The infecting organism was a rare strain of *Candida albicans* which causes vaginitis in yeasts but not in humans.

culations, and columns that show no significant variations, should be omitted.

Another very common, but useless, table is the word list. Table 5 is a typical example. This information could, quite obviously, be presented in the text. A good copy editor will kill this kind of table and incorporate the data into the text. I have done this myself thousands of times. Yet, when I have done it (and this leads to the next rule about tables), I have found more often than not that much or all of the information was already in the text. Thus, the rule: Present the data in the text, or in a table, or in a figure. *Never* present the same data in more than one way.

Tables 1 to 5 provide typical examples of the kinds of material that should not be tabulated. Now let us look at material that should be tabulated.

How to Arrange Tabular Material

Once having decided to tabulate, you must (or at least should) ask yourself exactly what should go into the table and in what form. Your first question is: "How do I arrange the data?" Since a table has both left-right and up-down dimensions, you have two choices. The data can be presented either horizontally or vertically. But *can* does not mean *should*; the data should be organized so that the like elements read *down*, not across.

Examine Tables 6 and 7. They are equivalent, except that Table 6 reads across, whereas Table 7 reads down. To use an old fishing expression, Table 6 is "bass ackward." Table 7 is the preferred format because (i) it allows the reader to grasp the information, and (ii) it is more compact and thus less expensive to print. The point about ease for the reader would seem to be obvious. (Did you ever try to add numbers that were presented horizontally rather than vertically?) The point about reduced printing costs refers to the fact that all columns must be wide or deep in the across arrangement, because of the diversity

TABLE 6. *Characteristics of antibiotic-producing Streptomyces*

Determination	S. fluoricolor	S. griseus	S. coelicolor	S. nocolor
Optimal growth temp (°C)	−10	24	28	92
Color of mycelium	Tan	Gray	Red	Purple
Antibiotic produced	Fluoricil-linmycin	Strepto-mycin	Rhol-monde-lay[a]	Nomycin
Yield of antibiotic (mg/ml)	4,108	78	2	0

[a] Pronounced "Rumley" by the British.

TABLE 7. *Characteristics of antibiotic-producing Streptomyces*

Organism	Optimal growth temp (°C)	Color of mycelium	Antibiotic produced	Yield of antibiotic (mg/ml)
S. fluoricolor	−10	Tan	Fluoricillinmycin	4,108
S. griseus	24	Gray	Streptomycin	78
S. coelicolor	28	Red	Rholmondelay	2
S. nocolor	92	Purple	Nomycin	0

TABLE 8. *Distribution of protein and ATPase in fractions of dialyzed membranes[a]*

Membranes from:	Fraction	ATPase U/mg of protein	Total U
Control	Depleted membrane	0.036	2.3
	Concentrated supernatant	0.134	4.82
E1 treated	Depleted membrane	0.034	1.98
	Concentrated supernatant	0.11	4.6

[a] Specific activities of ATPase of nondepleted membranes from control and treated bacteria were 0.21 and 0.20, respectively. Membranes were prepared from cells treated with colicin E1 as described in the legend to Fig. 4.

of elements, whereas some columns (especially those with numbers) can be narrow without runovers in the down arrangement. Thus, Table 7 appears to be smaller than Table 6, although it contains exactly the same information.

Table 8 is an example of a well-constructed table (reprinted from the "Instruction to Authors" of the *Journal of Bacteriology*). It reads down, not across. It has headings that are clear enough to make the meaning of the data understandable without reference to the text (a rule for a good table). It has explanatory footnotes, but it does not repeat excessive experimental detail. Note the distinction here. It is proper to provide enough information that the meaning of the data is apparent without reference to the text, but it is improper to provide *in the table* the experimental detail that would be required to repeat the experiment. The detailed materials and methods used to derive the data should remain in the section of that name.

Exponents in Table Headings

If possible, avoid using exponents in table headings. Confusion has resulted because some journals use positive exponents and some use negative exponents to mean the same thing. For example, the *Journal of Bacteriology* uses "cpm × 10^3" to refer to thousands of counts per minute, whereas *The Journal of Biological Chemistry* uses "cpm × 10^{-3}" for the same thousands of counts. If it is not possible to avoid such labels in table headings (or in figures), it may be worthwhile to state in a footnote (or the figure legend) what convention is being used, in words that eliminate the ambiguity.

Marginal Indicators

It is a good idea to identify in the margin of the text the location of the first reference to each table. Simply write "Table 3" (for example) in the margin and circle it. This procedure is a good check to make sure that you have indeed cited each table in the text, in numerical order. Mainly, however, this procedure provides flags so that the printer, at the page makeup stage (when galley proofs are converted to page proofs), will know where to break the text to insert the tables. If you do not mark location, a copy editor will; however, the copy editor might miss the first reference to a table, and the table could then be placed far from the primary text mention of it. Moreover, you might want to make passing reference to a table early in the paper but would prefer to have the table appear later in the paper. Only by your marginal notes will the copy editor and printer have access to your wishes in this regard.

Captions, Footnotes, and Abbreviations

The caption of the table (and the legend of a figure) is like the title of the paper itself. That is, the caption or legend should be concise and not divided into two or more clauses or sentences. Unnecessary words should be omitted.

Give careful thought to the footnotes to your tables. If abbreviations must be defined, you often can give all or most of the definitions in Table 1. Then later tables can carry the simple footnote: "Abbreviations as in Table 1."

Note that "temp" (Tables 1, 2, 6, and 7) is used as an abbreviation for "temperature." Because of space limitations in tables, almost all journals encourage abbreviation of certain words commonly used in tables. Such abbreviations would not be allowed in the text. Capitalize any such abbreviation used as the first word in a column heading; do not use periods (except after "no."). Get into the habit of using these abbreviations (Appendix 2) so that you and your typist can lay out tables properly. This is particularly helpful in designing camera-ready tables.

Camera-Ready Copy

These days, most authors in most institutions have access to word-processing equipment or, at least, modern electric typewriters equipped with carbon ribbons. Once you have learned how to design effective tables, you (your departmental secretary, really) can use this equipment to prepare *camera-ready* tables. More and more authors are doing this, either on their own or after being pushed by journal editors or managing editors. The advantages are substantial to the author, to the journal, and to the literature. A camera-ready table can be photographically reproduced in the journal. The advantage to you as the author is that you can escape the laborious chore of reading proof of the table. (The camera doesn't make typographical errors.) The advantage to the journal is that the cost of reproducing the table has been reduced because there is no need to keyboard the material, read proof, or make corrections. The advantage to the literature is that published data will contain fewer errors. Any errors in your original copy will of course remain, but the ubiquitous printer's errors of the past, to which tables were especially susceptible, can be totally avoided by submission of acceptable camera-ready copy.

Other parts of the manuscript can also benefit from the camera-ready approach. That way you will get what *you* want, not what a copy editor or printer thinks you want. Camera-ready copy works beautifully for complicated mathematical and physical formulas, chemical structures, portions of genetic maps, diagrams, and flow charts. Why not try it?

Chapter 13

How to Prepare
Effective Illustrations

A picture may instantly present what a book could set forth only in a hundred pages.

—IVAN SERGEYEVICH TURGENEV

When to Illustrate

In the previous chapter, I discussed certain types of data that should *not* be tabulated. They should *not* be turned into figures either. Basically, graphs are pictorial tables.

The point is this. Certain types of data, particularly of the sparse type or of the type that is monotonously repetitive, do not need to be brought together in either a table or a graph. The facts are still the same: The cost of preparing and printing an illustration is high, and we should consider illustrating our data only if the result is a real service to the reader.

This bears repeating, because many authors, especially those who are still beginners, think that a table, graph, or chart somehow adds importance to the data. Thus, in the search for credibility, there is a tendency to convert a few data elements into an impressive-looking graph or table. My advice is don't do it. Your more experienced peers and most journal editors will not be fooled; they will soon deduce that (for example) three of four curves in your graph are simply the standard conditions and that the meaning of the fourth curve could have been

55

stated in just a few words. Attempts to dress up scientific data are usually doomed to failure.

If there is only one curve on a graph, can you describe it in words? Possibly only one value is really significant, either a maximum or a minimum; the rest is window dressing. If you determined, for example, that the optimum pH value for a particular reaction was pH 8.1, it would probably be sufficient to state something like "Maximum yield was obtained at pH 8.1." If you determined that maximum growth of an organism occurred at 37°C, a simple statement to that effect is better economics and better science than a graph showing the same thing.

If the choice is not graph versus text but graph versus table, your choice might relate to whether you want to impart to readers exact numerical values or simply a picture of the trend or shape of the data. Rarely, there might be a reason to present the same data in both a table and a graph, the first presenting the exact values and the second showing a trend not otherwise apparent. (This procedure seems to be rather common in physics.) Most editors would resist this obvious redundancy, however, unless the reason for it was compelling.

An example of a nice, but unneeded, bar graph is shown in Fig. 1. This figure could be replaced by one sentence in the text: "Among the test group of 56 patients who were hospitalized for an average of 14 days, 6 acquired infections."

When is an illustration justified? There are no clear rules, but let us examine the types of illustrations that are in common use in scientific writing—graphs and photographs—with some indications for their effective use.

When to Use Graphs

Perhaps we should start with graphs (which are called line drawings in printing terminology) because they are very similar to tables as a means of presenting data in an organized way. In fact, the results of many experiments can be presented either as tables or as graphs. How do we decide which is preferable? This is often a difficult decision. A good rule might be this: If the data show pronounced trends, making an interesting picture, use a graph. If the numbers just sit there, with no exciting trend in evidence, a table should be satisfactory (and certainly easier and cheaper for you to prepare).

Examine Table 9 and Fig. 2, both of which record exactly the same data. Either format would be acceptable for publication, but I think Fig. 2 is clearly superior to Table 9. In the figure, the synergistic action of the two-drug combination is immediately apparent. Thus, the reader

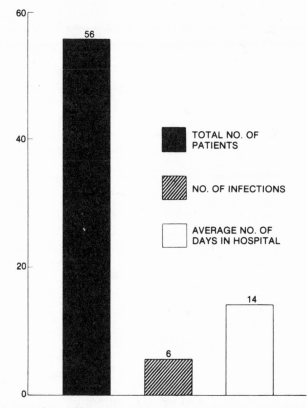

FIG. 1.   *Incidence of hospital-acquired infections. (Courtesy of Erwin F. Lessel.)*

can quickly grasp the significance of the data. It is also obvious in the graph that streptomycin is more effective than is isoniazid, although its action is somewhat slower; this aspect of the results is not readily apparent from the table.

TABLE 9.   *Effect of streptomycin, isoniazid, and streptomycin plus isoniazid on Mycobacterium tuberculosis[a]*

Treatment[b]	Percentage of negative cultures at:			
	2 wk	4 wk	6 wk	8 wk
Streptomycin	5	10	15	20
Isoniazid	8	12	15	15
Streptomycin + isoniazid	30	60	80	100

[a] The patient population was described in a preceding paper (61), although it has now become somewhat less patient.

[b] Highest quality available from our supplier (Town Pharmacy, Podunk, IA).

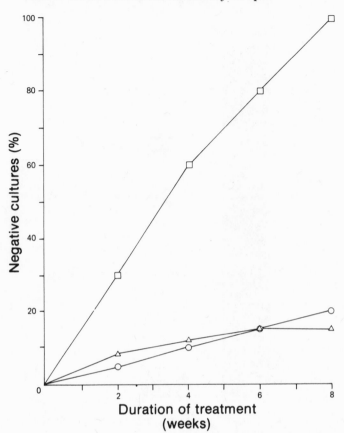

FIG. 2.   *Effect of streptomycin* (O), *isoniazid* (△), *and streptomycin plus isoniazid* (□) *on Mycobacterium tuberculosis. (Courtesy of Erwin F. Lessel.)*

How to Draw Graphs

Your first, hand-drawn graph will no doubt be prepared on lined graph paper. The final India ink drawing, however, should normally not be done on graph paper. Figure 3 shows the second-rate product that results from photographing and printing a graph submitted on typical graph paper. (One wonders also what "% cumulative" means. Some of us might guess that "mcg/mcl" means micrograms per microlitre; if the author's abbreviations were used correctly, they mean millicentigrams per millicentilitre, whatever that is.)

There is a type of graph paper printed with coordinate lines in light blue that is acceptable, because the light blue does not record when

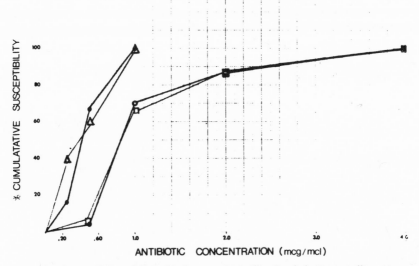

Fig. 3. *Susceptibility of clinical isolates to plentycillin. Symbols:* O, *penicillin-resistant Staphylococcus aureus;* Δ, *penicillin-susceptible S. aureus;* □, *S. epidermidis;* ●, *gross national product of Tanganyika in 1968.*

photographed. Blue tracing cloth is also sometimes used, but standard white tracing paper is readily available, inexpensive, and satisfactory. Let us note here, however, that it is *not* smart to submit original drawings to the journal. Such drawings are susceptible to damage during the review and printing processes, and of course they can be lost. It *is* smart to make photographic prints from the originals and to send the prints with the manuscript.

The easy way to prepare a graph is to draw in the data points and the curves, and then to add the numbers and abscissa-ordinate labels by typewriter. But no good journal will accept such a graph. The reason is obvious: typewritten letters are too small to withstand the photographic reduction to which graphs are normally subjected. Most journals have a two-column format and most graphs are sized to fit one column. The width of one column in a typical scientific journal (7 inches by 10 inches) is likely to be 16 picas, which is about 2⅝ inches (8.4 cm). Most typewritten graphs, when reduced to this width, are illegible. Therefore, the rule is "Absolutely no part of the graph or drawing should be typewritten" ("Instructions to Authors," *Journal of Bacteriology*). Some kind of lettering device must be used.

Even when a lettering set or transfer letters are being used, it is important to remember *why* they are being used. The size of the lettering

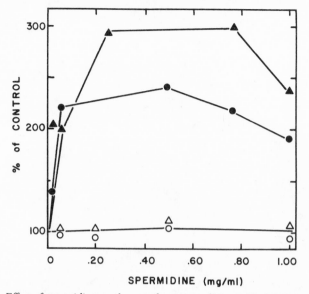

FIG. 4. *Effect of spermidine on the transformation of B. subtilis BR 151. Competent cells were incubated for 40 min with spermidine prior to the addition of 5 μg of donor DNA per ml (●) or 0.5 μg of donor DNA per ml (▲). DNA samples of 5 μg (○) or 0.5 μg per ml (△) were incubated for 20 min prior to the addition of cells. (Molec. Gen. Genet. 178:21–25, 1980; courtesy of Franklin Leach.)*

must be based on the anticipated photographic reduction that will occur in the printing process. This factor becomes especially important if you are combining two or more graphs into a single illustration.

Figure 4 is a rather nice graph. The lettering was large enough to withstand photographic reduction. It is boxed in rather than two-sided (compare with Fig. 2), making it a bit easier to estimate the values on the right-hand side of the graph. The scribe marks point inward rather than outward.

Size and Arrangement of Graphs

Examine Fig. 5. Obviously, the lettering was not large enough to withstand the reduction which occurred, and most readers would have difficulty in reading the ordinate and abscissa labels. Actually, Fig. 5 effectively illustrates two points. First, the lettering must be of sufficient size to withstand reduction to column or page width. Second, because width is the important element from the printer's point of view, it is often advisable to combine figures "over and under" rather than "side by side." If the three parts of Fig. 5 had been prepared in the "over and under" arrangement, the photographic reduction would have been nowhere near as drastic and the labels would be much more readable.

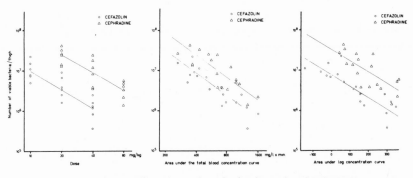

FIG. 5.   *Dose-effect relationship of cefazolin and cephradine (35).*

The spatial arrangement of Fig. 5 may not be ideal, but the combination of three graphs into one composite arrangement is entirely proper. Whenever figures are related and can be combined into a composite, they should be combined. The composite arrangement saves space and thus reduces printing expense. More importantly, the reader gets a much better picture by seeing the related elements in juxtaposition.

Do not extend the ordinate or the abscissa (or the explanatory lettering) beyond what the graph demands. For example, if your data points range between 0 and 78, your topmost index number should be 80. You might feel a tendency to extend the graph to 100, a nice round number; this urge is especially difficult to resist if the data points are percentages, for which the natural range is 0 to 100. You must resist this urge, however. If you do not, parts of your graph will be empty; worse, the live part of your graph will then be restricted in dimension, because you have wasted perhaps 20% or more of the width (or height) with empty white space.

In the example above (data points ranging from 0 to 78), your lettered reference numbers should be 0, 20, 40, 60, and 80. You should use short index lines at each of these numbers and also at the intermediate 10s (10, 30, 50, 70). Obviously, a reference stub line between 0 and 20 could only be 10. Thus, you need not letter the 10s, and you can then use larger lettering for the 20s, without squeezing. By using such techniques, you can make graphs simple and effective instead of cluttered and confusing.

Symbols and Legends

If there is space in the graph itself, use it to present the key to the symbols. In the bar graph (Fig. 1), the shadings of the bars would have been a bit difficult to define in the legend; given as a key, they need no further definition (and additional typesetting, proofreading, and expense are avoided).

If you must define the symbols in the figure legend, you should use only those symbols that are considered standard, for which the printer will have type. Perhaps the most standard symbols are open and closed circles, triangles, and squares (O, △, □, ●, ▲, ■). If you have just one curve, use open circles for the reference points; use open triangles for the second, open squares for the third, closed circles for the fourth, and so on. If you need more symbols, you probably have too many curves for one graph and you should consider dividing it into two. If you must use a few more symbols, every printer has the multiplication sign (×), and many printers have (or could construct) half symbols (◐, ◑, ◭, etc.). Different types of connecting lines (solid, dashed) can also be used.

Graphs must be neatly drawn. In printing, these "line shots" come out black and white; there are no grays. Anything drawn too lightly (plus most smudges and erasures) will not show up at all in printing; however, what does show up may show up very black, perhaps embarrassingly so. Fortunately, you can usually determine in advance what your printed graphs will look like, simply by making photocopies. Most office photocopiers seem to act like printers' cameras.

What I have said above, about the use of lettering sets, etc., assumes that you will make the graphs yourself. If so, these directions may be useful. If someone else in your institution prepares the graphs, you may be able to provide reasonable instructions if you are aware of the essential elements. If you are not experienced in graph-making, and such talent is not readily available in your institution, you should probably try to find a good commercial art establishment. Scientists are sometimes surprised that a commercial artist can do in minutes, at reasonable cost (usually), what it would take them hours to do. Graph-making is not a job for amateurs.

As to the legends, they should *always* be typed on a separate page, never at the bottom or top of the illustrations themselves. The main reason for this is that the two portions must be separated in the printing process, the legends being produced by typesetting and the illustrations by photographic processes.

Photographs and Micrographs

If your paper is to be illustrated with one or more photographs, which become halftones in the printing process, there are several factors to keep in mind. As with graphs, the size (especially width) of the print in relation to the column and page width of the journal is extremely important. Thus, size should be important to you in making your material fit the journal page. It is important to the journal because the costs of halftone reproduction are very high. Authors may have to worry more about the costs in the future, as more and more journals require authors to pay part of the costs. Some time ago, *The Journal of Biological Chemistry* introduced a charge of $45 for each normal halftone plate and a charge of $100 per plate for electron micrographs.

The most important factor to worry about, however, is a proper appreciation of the *value* of the photographs for the story you are presenting. The value can range from essentially zero (in which case, like useless tables and graphs, they should not be submitted) to a value that transcends the text itself. In many studies of cell ultrastructure, for example, the significance of the paper lies in the photographs.

If your photographs (especially electron micrographs) are of prime importance, you should first ask yourself which journal has high-quality reproduction standards (halftone screens of 150 to 200 lines, coated stock) for printing fine-structure studies. In biology, the journals published by the American Society for Microbiology and by The Rockefeller University Press are especially noted for their high standards in this respect.

Cropping and Framing

Whatever the quality of your photographs, you want to have them printed legibly. To some degree, you can control this process yourself if you use your head.

If you are afraid that detail might be lost by excessive reduction, there are several ways you might help. Usually, you should put crop marks on the margins of the photographs. Seldom do you need the whole photograph, right out to all four edges. Therefore, frame the important part; this is especially useful if you can frame the width to the column or page width of the journal. You can then boldly write on the edge of the print or on the cover sheet: "Print one-column width (or page width) without photographic reduction." Dealing with such a carefully cropped photograph containing a reasonable instruction from the author, most copy editors will be pleased to oblige. Figures 6, 7, and 8 show photographs with and without cropping. Figure 6 indicates how cropping can be used to decrease the dimensions of the photograph. Figures 7 and 8 show how cropping can be used to enlarge an area of a photograph. The greatest fidelity of reproduction results when you furnish exact-size photographs, requiring neither enlargement nor reduction. Significant reduction (more than 50%) should be avoided. (Greater reduction of graphs is all right, if the lettering can withstand it.) There is no need for

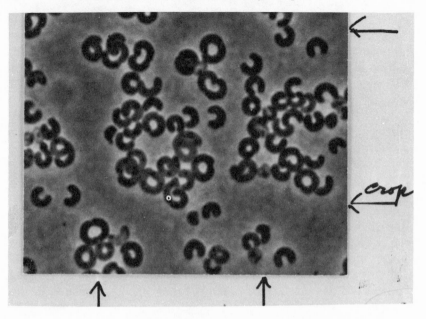

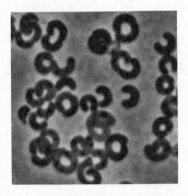

FIG. 6. *Cell forms of Rhodocyclus purpureus. Phase contrast. Magnification, ×3,000. Original photograph (top). Cropped version (bottom). (Reprinted from Int. J. Syst. Bacteriol. 28:285, 1978.)*

"glossy" prints, as requested by some journals, provided the matte surface is smooth.

You should *never* put crop marks directly on a photograph. Margin marks can sometimes be used, especially if the photographs are mounted on Bristol board or some other backing material. Otherwise, crop marks may be placed on a tracing paper overlay.

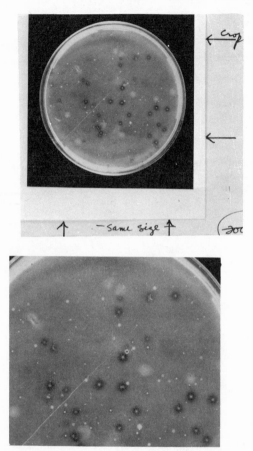

FIG. 7. *Soil dilution (0.1 ml) spread on a yeast cell agar plate. Photographed after 3 days of incubation at 26°C. Note the lytic zones around some of the colonies. Uncropped (top); cropped (bottom). (Reprinted from Int. J. Syst. Bacteriol. 28:368, 1978.)*

A useful trick you might try is as follows: Cut two "Ls," perhaps 6 inches high, 3 inches at the base, and 1 inch wide, from black construction paper. If you now invert one "L" and place it over the other, you have at your disposal an adjustable rectangle with which to frame your photographs. By such "framing," you can place crop marks where they give you the best picture.

Necessary Keys and Guides

If you can't crop down to the features of special interest, consider superimposing arrows or letters on the photographs. In this way, you

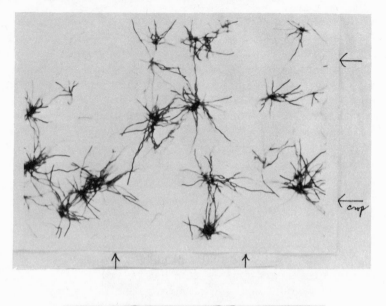

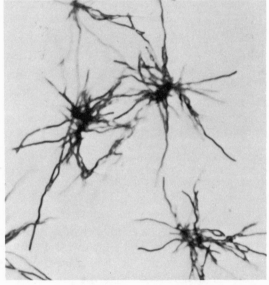

FIG. 8. *Sporulating aerial mycelium of a strain of Actinomycetales. Uncropped (top); cropped (bottom). (Reprinted from Int. J. Syst. Bacteriol. 28:56, 1978.)*

can draw the reader's attention to the significant features, while making it easy to construct meaningful legends.

Always mark "top" on what you consider to be the top of the photograph. (Mark it on the back, with a soft pencil.) Otherwise, the photograph (unless it has a very obvious top) may be printed upside-down or sideways. If the photograph is of a field which can be printed in any orientation, mark "top" on a narrow side. (That is, on a 4 × 6 or 8 × 10 print, the 4-inch or 8-inch dimension should be the width, so that less reduction will be required to reach one-column width or page width.)

As with tables, it is a good idea to indicate the preferred location for each illustration. In this way, you will be sure that all illustrations have been referred to in the text, in one-two-three order, and the printer will know how to weave the illustrations into the text so that each one is close to the text related to it.

With electron micrographs, put a micrometer marker directly on the micrograph. In this way, regardless of the percentage of reduction (or even enlargement) in the printing process, the magnification factor is clearly in evidence. The practice of putting the magnification in the legend (e.g., ×50,000) is not advisable and some journals no longer allow it, precisely because the size (and thus the magnification) is likely to change in printing. And, usually, the author forgets to change the magnification at the proof stage.

Color Photographs

Although many laboratories are now equipped to make color photographs, it is rare that color photographs are printed in journals; the cost is prohibitive. To print in full color, a negative and a plate have to be made for each of the four colors (three primary colors plus black), and the sheet containing the color illustrations may have to be printed on special presses. Some journals will print a color illustration if the editor agrees that color is necessary to show the particular phenomenon and if the author can pay (perhaps from grant funds) part or all of the additional printing cost (typically $1,500 to $2,000 for a color plate). Therefore, your laboratory photography should normally be done in black and white because that is what can be printed. Although color photographs can be printed in black and white, they often wash out and do not have the fidelity of original black-and-white photographs.

Pen-and-Ink Illustrations

In some areas (especially descriptive biology), pen-and-ink illustrations (line drawings) are superior to photographs in showing significant details. Such illustrations are also common in medicine, especially in presenting anatomic views, and indeed have become virtually an art form. Normally, the services of a professional illustrator are required when such illustrations are deemed necessary.

Chapter 14

How to Type
the Manuscript

*Then the black-bright, smooth-running, clicking
clean
Brushed, oiled and dainty typewriting machine,
With tins of ribbons waiting for the blows
Which soon will hammer them to verse and prose.*
                                    —JOHN MASEFIELD

Importance of Clean Typing

When you have finished the experiments and written up the work, the final typing of the manuscript is not important because, if your work is good, sound science, it will be accepted for publication. Right? I have news for you, friend. That is *wrong.* A badly typed manuscript will not only fail to be accepted for publication; in most journal operations, a sloppily prepared manuscript will not even be *considered.*

At the Publications Office of the American Society for Microbiology, which is not atypical in this respect, every newly submitted manuscript is examined *first* simply on the basis of the *typing.* As an irreducible minimum, the manuscript must be typed (not handwritten), double-spaced (not single-spaced), on one side of the sheet only (not both sides); two complete copies (including two sets of tables, graphs, and photographs) must be provided; and reasonable adherence to the style of the journal (appropriate headings, proper form of literature citation, presence of a heading abstract) must be in evidence. If the manuscript fails on any of these major points, it is immediately returned to the author.

Consider this a cardinal rule: *Before* the final typing of your manuscript, carefully reexamine the "Instructions to Authors" of the journal to which you are submitting the manuscript. Also look carefully at a recent issue of that journal. Pay particular attention to those aspects of editorial style which tend to vary widely from journal to journal, such as the style of literature citation, headings and subheadings, size and placement of the abstract, design of tables and figures, and treatment of footnotes.

By the way, an increasing number of journals seem to be refusing to accept footnotes. The main reason for this is the significant printing cost of carrying the footnotes at the bottom of the page, in a different font of type. Furthermore, footnotes are disruptive to readers, making papers more difficult to read quickly with comprehension. Therefore, do not use footnotes unless a particular journal requires them for some purpose. (Most journals require "present address" footnotes if an author has moved; some journals require that the names of manufactured products be footnoted, with the footnotes giving the names and addresses of the manufacturers.) Whenever somewhat extraneous material needs to be mentioned, do it parenthetically in the text. Some journals have a "References and Notes" section at the end of each paper, thus obviating the need for text footnotes. Avoidance of footnotes is encouraged for most kinds of writing, and it is strongly encouraged for the writing of scientific papers.

In an ideal world, perhaps good science could be published without regard to the format of the carrier (the typed manuscript). In the real world, however, busy editors and reviewers, who serve without salary in most operations, simply cannot and will not take the time to deal with messy, incomplete manuscripts. Further, most experienced editors believe that there is a direct relationship involved; a poorly prepared manuscript is, almost without fail, the carrier vehicle of poor science.

Therefore, my advice to you is firm on this point. If you want your manuscript to be published (and why else would you be submitting it?), make very sure that the submitted manuscript is typed neatly, without errors, in the style of the journal, and that it is complete in all respects. *This is a must.*

Paging the Manuscript

It is advisable to start each section of a manuscript on a new page. The title and authors' names are usually on the first page and this page should be number 1. The Abstract is on the second page. The Introduction starts on the third page, and each succeeding section (Materials and Methods, Results, etc.) then starts on a new page. The Literature Cited starts on a new page. Each table is on its own page. Figure legends are grouped on one separate page. The tables and figures (and figure legends) should be assembled at the back of the manuscript, not interspersed through it.

Historically, the "new page" system was a requirement of many journals because the older typesetting technology required separation of different material. If, for example, the journal style called for 8-point type in the Abstract and 9-point type in the Introduction, these two sections had to go to different lead-casting machines. Thus, the copy had to be cut unless the natural divisions were provided for in advance.

Because of the flexibility of modern phototypesetters, copy no longer has to be cut. Yet, it is still a good idea to preserve these natural divisions. Even if the divisions do not aid the printing process, they often are useful to you in the manuscript revision process. Often, for example, you may decide (or the reviewers may decree) that a particular method should be added, expanded, contracted, or deleted. The chances are that the Materials and Methods section could be retyped, from the page of the change to the end, without disturbing the rest of the manuscript. Probably only the amount of white space on the last page of Materials and Methods will change. Even if the new material requires additional space, you need not disturb the later sections. Suppose, for example, that the Materials and Methods section in your original manuscript concludes on page 5, the Results begin on page 6, and there isn't enough white space on page 5 to allow for insertion of the needed new material. Simply retype Materials and Methods from the page of change on, going from page 5 to page 5a (and 5b, etc., if necessary). The Results and later sections need not be touched.

Margins and Headings

Your manuscript should have wide margins. A full inch (ca. 2.5 cm) at the top, bottom, and both sides is about minimum. You will need this space yourself during revisions of the manuscript. Later, the copy editor and the printer will need this space to enter necessary instructions. Also, it is advantageous to use paper with numbered lines, for ease in pointing to problems throughout the editorial and printing process.

Before the final typing, examine your headings carefully. The main headings ("Materials and Methods," etc.) are usually no problem. These headings should be centered, with space above and below.

In addition to main headings, most journals use subheadings (e.g., boldface paragraph lead-ins). These should be designed as convenient signposts to help direct the reader through the paper. Consult a recent issue of the journal to determine what kinds of headings it uses. If the journal uses boldface lead-ins, have them typed that way. You or your typist can add wavy underlines (to denote boldface), so that the copy editor and printer will have no doubt as to where and what your headings are. If the journal uses italic lead-ins, have the typist type a single underline under your lead-in subheadings.

Do not make the common mistake of using a third (or even a fourth) level of heading, unless such usage is specified by the journal. Two levels of headings are usually sufficient for research papers, and many journals do not permit more. Review journals, however, usually specify three or four levels of headings because of the greater length of review papers. A good discussion of the use of "topical heads" is given by Skillin et al. (36).

Special Problems

Keep in mind that the keyboarding done by your typist is not very different from that done later by the printer. If your typist has a problem with your manuscript, it is likely that the printer will also have a problem. See if you can identify and then resolve some of these problems, to make it easy on your typist and the printer. For example, most typesetting devices (like the office typewriter) move relentlessly forward, meaning that it is difficult or impossible to set certain overs and unders. Thus, an over-under fraction, such as $\dfrac{ab - c}{de - x}$, is a real problem. Change the form to $(ab - c)/(de - x)$ and there is no problem. Likewise, it is difficult (and in Monotype, for example, impossible) to set an inferior letter directly under a superior. Thus, $a_2{}^1$ is not a problem but a_2^1 is a problem. The term $\sqrt{ax^2}$ in the text is a real problem for almost all typesetting devices. The easy alternative is to state "the square root of ax^2." If a formula simply cannot be put in a form suitable for keyboarding, you should consider furnishing it as an India ink drawing. You will thus save your typist and the typesetter a lot of trouble, and you might save yourself a lot of grief. The camera will set your formula perfectly; the typesetter might not.

There is American-English and there is English-English. To avoid problems for yourself as well as for typesetters and proofreaders, use American spellings in a manuscript being submitted to a journal in the

United States, and use British spellings in a manuscript being submitted to a journal in Great Britain.

Final Review

One final note. After the manuscript has been typed, you would be wise to do two things.

First, read it yourself. You would be surprised how many manuscripts are submitted to journals without being proofread after final typing—manuscripts so full of typing errors that sometimes even the author's name is misspelled. Recently, a manuscript was submitted by an author who was too busy to proofread not only the final typing of the manuscript but also the covering letter. His letter read: "I hope you will find this manuscript exceptable." We did.

Second, ask one or more of your colleagues to read your manuscript before you submit it to a journal (and perhaps before final typing). It may well be that the meaning of one or more parts of your paper is completely unclear to your colleague. Of course, this may be because your colleague is dense, but it is just possible that this portion of your manuscript is not as clear as it could be. You might also ask a scientist working in a different field to read your paper and to point out words and phrases he or she doesn't understand. This is perhaps the easiest way to identify the jargon that may be present in your manuscript.

Expect to sweat a bit, if you haven't already done so. As the "Instructions to Authors" of the *Journal of General Microbiology* once put it, "Easy reading is curst hard writing."

Chapter 15

Where and How to Submit the Manuscript

I keep six honest serving-men
(They taught me all I knew);
Their names are What and Why and When
And How and Where and Who.
—RUDYARD KIPLING

Choosing the Journal

The choices of where and how to submit the manuscript are important. Some manuscripts are buried in inappropriate journals. Others are lost, damaged, or badly delayed because of carelessness on the part of the author.

The first problem is where to submit the manuscript. (Actually, you will have already reached a decision on this point *before* the final typing of the manuscript.) Obviously, your choice depends on the nature of your work; you must identify those journals which publish in that subject area. A good way to get started or to refresh your memory is to scan a recent issue of *Current Contents*. It is usually easy to determine, on the basis of journal titles alone, which journals might publish papers in your field. Only by examination of the tables of contents, however, can you determine which journals *are* publishing papers in your field.

Another easy way to get started is to scan *Journal Citation Reports* (an annual volume supplementing the *Science Citation Index*). This bibliographic tool makes it possible to study a number of important aspects of the use made of individual journals.

To identify which journals might publish your manuscript, you should do several things: Read the masthead statement in a current issue of each journal you are considering; read the "scope" paragraphs that are usually provided in the "Instructions to Authors"; and look carefully at the table of contents of a current issue.

Because journals have become more specialized, and because even the older journals have changed their scope frequently (of necessity, as science itself has changed), you must make sure that the journal you are considering is currently publishing work of the kind you propose to submit.

If you submit your manuscript to a wrong journal, one of three things can happen, all bad.

First, your manuscript will simply be returned to you, with the comment that your work "is not suitable for this journal." Often, however, this judgment is not made until *after* review of the manuscript. A "not suitable" notice after weeks or months of delay is not likely to make you happy.

Second, if the journal is borderline in relation to your work, your manuscript may receive poor or unfair review, because the reviewers (and editors) of that journal may be only vaguely familiar with your specialty area. You may be subjected to the trauma of rejection, even though the manuscript would be acceptable to the right journal. Or you could end up with a hassle over suggested revisions, which you do not agree with and which do not improve your manuscript. And, if your manuscript really does have deficiencies, you will not be able to benefit from the sound criticism that would come from the editors of the right journal.

Third, even if your paper is accepted and published, your glee will be short-lived if you later find that your work is virtually unknown because it is buried in a publication that your peers do not read.

The Prestige Factor

If several journals are right, does it matter which you select? Perhaps it shouldn't, but it does. There is the matter of *prestige*. It may be that your future progress (promotions, grants) will be determined solely by the numbers game. But not necessarily. It may well be that a wise old bird sitting on the faculty committee or the grant review panel will recognize and appreciate quality factors. A paper published in a "garbage" journal simply does not equal a paper published in a prestigious journal. In fact, the wise old bird (and there are quite a few around in science) may be more impressed by the candidate with one or two solid publications in prestigious journals than by the candidate with ten or more publications in second-rate journals.

How do you tell the difference? It isn't easy, and of course there are many gradations. In general, however, you can form reasonable judgments by just a bit of bibliographic research. You will certainly know the important papers that have recently been published in your field. Make it your business to determine *where* they were published. If most of the real contributions to your field were published in Journal A, Journal B, and Journal C, you should probably limit your choices to those three journals. Journals D, E, and F, upon inspection, contain only the lightweight papers, so they could be eliminated as your first choice, even though the scope is right.

You may then choose among Journals A, B, and C. Suppose that Journal A is a new, attractive journal published by a commercial publisher as a commercial venture, with no sponsorship by a society or other organization; Journal B is an old, well-known, small journal, published by a famous hospital or museum; and Journal C is a large journal published by the principal scientific society representing your field. As a general rule (although there are many exceptions), Journal C (the society journal) is probably the most prestigious. It also will have the largest circulation (partly because of quality factors, partly because society journals are less expensive than most others, at least to society members). By publication in such a journal, your paper may have its best chance to make an impact on the community of scholars you are aiming at. Journal B might have almost equal prestige, but it might have a very limited circulation, which would be a minus; it might also be very difficult to get into, if most of its space is reserved for in-house material. Journal A (the commercial journal) almost certainly has the disadvantage of low circulation (because of its comparatively high price, which is the result of both the profit aspect of the publisher and the fact that it does not have the backing of a society or institution, with a built-in subscription list). Publication in such a journal may result in a somewhat restricted distribution for your paper.

The Circulation Factor

If you want to determine the comparative circulation of several journals, there is an easy and accurate way to do it for U.S. journals. Look among the last few pages of the November and December issues, and you will find a "Statement of Ownership, Management and Circulation." The U.S. Postal Service requires that each publisher granted second-class mailing privileges (and almost all scientific journals qualify) file and publish an annual statement. This statement must include basic circulation data.

If you can't determine the comparative circulation of journals you are considering and have no other way of assessing comparative prestige factors, a very useful tool exists for rating scientific journals. I refer to *Journal Citation Reports.* By use of this reference document, you can determine which journals are cited most frequently, both in gross quantitative terms and in terms of average citations per article published ("impact factor"). The impact factor especially seems to be a very reasonable basis for judging the quality of journals. If the average paper in Journal A is cited twice as frequently as the average paper in Journal B, there is little reason to question that Journal A is the more important journal.

The Frequency Factor

Another factor to consider is frequency of the journal. The "publication lag" of a monthly journal is almost always shorter than that of a quarterly journal. Assuming equivalent review times, the additional delay of the quarterly will range up to 2 or 3 months. And, since the publication lag, including the time of editorial review, of many (probably most) monthlies ranges between 4 and 7 months, the lag of the quarterly is likely to run up to 10 months. Remember, also, that many journals, whether monthly, bimonthly, or quarterly, have backlogs. It sometimes helps to ask colleagues what their experience has been with the journal(s) you are considering. If the journal publishes "Received for Publication" dates, you can figure out for yourself what the average lag time is.

Be wary of new journals, especially those not sponsored by a society. The circulation may be minuscule and the journal might fail before it, and your paper, become known to the scientific world.

Packaging and Mailing

After you have decided where to submit your manuscript, do not neglect the nitty gritty of sending it in.

How do you wrap it? *Carefully.* Take it from a long-time managing editor: *Many* manuscripts are lost, badly delayed, or damaged in the mail, often because of improper packaging. (Special note: *Always* retain at least one copy of the manuscript. I have known of several dummies who mailed out the only existing copies of their manuscripts, and there was an unforgettable gnashing of teeth when the manuscripts were forever lost.)

Use a strong manila envelope or even a reinforced mailing bag. Whether or not it is a clasp envelope, you would be wise to put a piece of reinforced tape over the sealed end.

Make *sure* that you apply sufficient postage and that you send the package by first-class mail. Much of the manila-envelope mail handled by the U.S. Postal Service is third-class mail, and your manuscript will be treated as third-class mail and delivered next month if you neglect to indicate "First Class Mail" clearly on the package or if you apply insufficient postage.

Most scientific journals do not require that authors supply stamped, self-addressed return envelopes, although most journals in other scholarly fields do enforce such a requirement. Apparently, the comparative brevity of scientific manuscripts makes it cost-effective for publishers to pay return postage rather than store many bulky envelopes.

Increasingly, both publishers and authors (their institutions) in the U.S. are using United Parcel Service. Many users are convinced that UPS is usually a bit faster and more reliable than the Postal Service.

Overseas mail should be sent *airmail*. A manuscript sent from Europe to the U.S., or vice versa, will arrive within 3 to 5 days if sent by airmail; by surface mail, the elapsed time will be 4 to 6 weeks.

The Covering Letter

Finally, it is worth noting that you should always send a covering letter with the manuscript. Manuscripts without covering letters pose immediate problems: To which journal is the manuscript being submitted? Is it a new manuscript, a revision requested by an editor (and, if so, which editor?), or is it a manuscript perhaps misdirected by a reviewer or an editor? If there are several authors, which one should be considered the submitting author, at which address? The address is of special importance, because the address shown on the manuscript is often not the current address of the contributing author.

Be kind to the editor and state why you have submitted that particular package. You might even choose to say something nice, as was done recently in a letter in impeccable English but written by someone whose native tongue was not English. The letter read: "We would be glad if our manuscript would give you complete satisfaction."

Sample Covering Letter

Dear Dr.————:

Enclosed are two complete copies of a manuscript by Mary Q. Smith and John L. Jones titled "Fatty Acid Metabolism in *Cedecia*

neteri." This manuscript is being submitted for possible publication in the Physiology and Metabolism section of the *Journal of Bacteriology.*

This manuscript is a new manuscript and it is not being considered elsewhere. The manuscript reports new findings that extend results we reported earlier in *The Journal of Biological Chemistry* (135:112–117, 1982). An abstract of this manuscript was presented earlier (Abstr. Annu. Meet. Am. Soc. Microbiol., p. 406, 1982).

Correspondence regarding this manuscript should be addressed to me at the address shown in the above letterhead (not the address shown on the manuscript, from which laboratory I have recently moved).

Sincerely,

Mary Q. Smith

Follow-up Correspondence

Most journals send out an "acknowledgment of receipt" form letter when the manuscript is received. If you know that the journal does not, attach a self-addressed postcard to the manuscript, so that the editor can acknowledge receipt. If you do not receive an acknowledgment in 2 weeks, call or write the editorial office to verify that your manuscript was indeed received. I know of one author whose manuscript was lost in the mail, and it was not until 9 months later that the problem was brought to light by his meek inquiry as to whether the reviewers had reached a decision about the manuscript.

The mails being what they are, and busy editors and reviewers being what they are, do not be concerned if you do not receive a decision within one month after submission of the manuscript. Most journal editors, at least the good ones, try to reach a decision within 4 to 6 weeks or, if there is to be further delay for some reason, provide some explanation to the author. If you have had no word about the disposition of your manuscript after 6 weeks have elapsed, it is not at all inappropriate to send a courteous inquiry to the editorial office. If no reply is received and the elapsed time becomes 2 months, a personal phone call to the editor is not out of place.

Chapter 16

The Review Process (How to Deal with Editors)

We realize that with all this help, there's a certain amount of arbitrariness in the decisions we make. But if we didn't have a system like this, we'd have to invent it. Otherwise we'd drown in a sea of unverified, incomplete, premature, half-baked work.

—ARNOLD S. RELMAN

Functions of Editors and Managing Editors

Editors and managing editors have impossible jobs. What makes their work impossible is the attitude of authors. This attitude was well expressed by Earl H. Wood of the Mayo Clinic in his contribution to a panel on the subject "What the Author Expects from the Editor." Dr. Wood said, "I expect the editor to accept all my papers, accept them as they are submitted, and publish them promptly. I also expect him to scrutinize all other papers with the utmost care, especially those of my competitors."

Somebody once said, "Editors are, in my opinion, a low form of life—inferior to the viruses and only slightly above academic deans."

And then there is the story about the Pope and the editor who died and arrived in heaven simultaneously. They were subjected to the usual initial processing and then assigned to their heavenly quarters. The Pope looked around his apartment, and found it to be spartan indeed. The editor, on the other hand, was assigned to a magnificent apartment, with plush furniture, deep pile carpets, and superb appointments. When the Pope saw this, he went to God and said: "Perhaps there has been a mistake. I am the Pope and I have been assigned to shabby quarters, whereas this lowly editor has been assigned to a lovely apartment." God answered: "Well, in my opinion there isn't anything very special about you. We've admitted over 200 Popes in the last 2,000 years. But this is the very first editor who ever made it to heaven."

Going back to the first sentence of this chapter, let us distinguish between "editors" and "managing editors." Authors should know the difference, if for no other reason than knowing whom to complain to when things go wrong.

An *editor* (some journals have several) decides whether to accept or reject manuscripts. Thus, the editor of a scientific journal is a scientist, often of preeminent standing. The editor not only makes the final "accept" and "reject" decisions, but also designates the peer reviewers upon whom he or she relies for advice. Whenever you have reason to object to the quality of the reviews of your paper (or the decision reached), your complaint should be directed to the editor.

It has been said that the role of the editor is to separate the wheat from the chaff and then to make sure that the chaff gets printed.

The *managing editor* is normally a full-time paid professional, whereas editors usually are unpaid volunteer scientists. (A few of the very large scientific and medical journals do have full-time paid editors. A number of other journals, especially those published commercially, pay salaries to their part-time editors.) Normally, the managing editor is not directly involved with the "accept-reject" decisions. Instead, the managing editor attempts to relieve the editor of all clerical and administrative detail in the review process, and is responsible for the later events that convert accepted manuscripts into published papers. Thus, when problems occur at the proof and publication stages, you should communicate with the managing editor.

In short, pre-acceptance problems are normally within the province of the editor, whereas post-acceptance problems are within the bailiwick of the managing editor. However, from my years of experience as a managing editor, I can tell you that there seems to be one fundamental law that everybody subscribes to: "Whenever anything goes wrong, blame the managing editor."

The Review Process

You, as an author, should have some idea of the whys and wherefores of the review process. Therefore, I will describe the policies and procedures which are typical in most editorial offices. If you can understand

(and perhaps even appreciate) some of the reasons for the editorial decisions that are made, perhaps in time you can improve the acceptance rate of your manuscripts, simply by knowing how to deal with editors.

When your manuscript first arrives at the journal editorial office, the editor (or the managing editor, if the journal has one) makes several preliminary decisions. First, is the manuscript concerned with a subject area covered by the scope of the journal? If it clearly is not, the manuscript is immediately returned to the submitting author, along with a short statement pointing to the reason for the action. Seldom would an author be able to challenge such a decision successfully, and it is usually pointless to try. It is an important part of the editor's job to define the scope of the journal, and editors I have known seldom take kindly to suggestions by authors, no matter how politely the comments are phrased, that the editor is somehow incapable of defining the basic character of his or her journal. Remember, however, that such a decision is not rejection of your data or conclusions. Your course of action is obvious: Try another journal.

Second, if the subject of the manuscript is appropriate for consideration, is the manuscript itself in suitable form for consideration? Are there two double-spaced copies of the manuscript? (Some journals require three.) Are they complete, with no pages, tables, or figures missing from either copy of the manuscript? Is the manuscript in the editorial style of the journal, at least as to the basics? If the answer to any of the above questions is "no," the manuscript is immediately returned to the author. Most journal editors will not waste the time of their valued editorial board members and consultants by sending poorly prepared manuscripts to them for review.

I know of one editor, a kindly man by nature, who became totally exasperated when a poorly prepared manuscript that was returned to the author was resubmitted to the journal with very little change. The editor then wrote the following letter, which I am pleased to print here as a warning to all students of the sciences everywhere:

> Dear Dr. _____:
> I refer to your manuscript "_____"
> and have noted in your letter of August 23 that you apologize without excuse for the condition of the original submission. There is really no excuse for the rubbish that you have sent forward in the resubmission.
>
> The manuscript is herewith returned to you. We suggest that you find another journal.
>
> Yours sincerely,
>
> _____

Only after these two preconditions (a proper manuscript on a proper subject) have been met is the editor ready to consider the manuscript for publication.

At this point, the editor must perform two very important functions. First, the basic housekeeping must be done. That is, careful records should be established so that both copies of the manuscript can be followed throughout the review process and (if the manuscript is accepted) the printing process. If the journal has a managing editor, and most of the large ones do, this activity is normally a part of his or her assignment. It is important that this work be done accurately, so that the whereabouts of manuscripts are known at all times. It is also important that the system include a number of built-in signaling devices, so that the inevitable delays in review, loss in the mails, and other mini and maxi disasters can be brought to the attention of the editor or managing editor at an early time.

Second, the editor must decide who will review the manuscript. In most journal operations, two reviewers are selected for each manuscript. Obviously, the reviewers must be peers of the author, or their recommendations will be valueless. Normally, the editor starts with the Editorial Board of the journal. Who on the Board has the appropriate subject expertise to evaluate a particular manuscript? Often, because of the highly specialized character of modern science, only one member (or no member) of the Board has the requisite familiarity with the subject of a particular manuscript. The editor must then obtain one or both reviews from non-Board members, often called "ad hoc reviewers" or "editorial consultants." (A few journals do not have Editorial Boards and depend entirely on ad hoc referees.) Sometimes, the editor must do a good bit of calling around before appropriate reviewers for a given manuscript can be identified.

What does a reviewer do, and why? What are the ethics and philosophy of peer review? I believe that these questions are well answered in "Guidelines for Reviewers, American Society for Microbiology," which is reprinted at the end of this chapter. The ASM guidelines are patterned after policies recommended by a committee of the Council of Biology Editors (CBE) some years ago. The same policies, or variations of them, have been adopted by many journals. Some of these policies were discussed at some length by DeBakey (18), a former Chairman of the CBE committee. It is important for both authors and reviewers to understand the procedural and policy framework that surrounds the peer review system.

Most journals use anonymous reviewers. A few journals make the authors anonymous by deleting their names from the copies of manuscripts sent to reviewers. My own experience is in accord with that of the distinguished Canadian scientist, J. A. Morrison, who has said (27): "It is occasionally argued that, to ensure fairness, authors should also be anonymous, even though that would be very difficult to arrange. Actually, editors encounter very few instances of unfairness and blatant bias expressed by referees; perhaps for 0.1 per cent or less of the manuscripts handled, an editor is obliged to discount the referee's comments."

If the reviewers have been chosen wisely, the reviews will be meaningful and the editor will be in a good position to arrive at a decision regarding publication of the manuscript. When the reviewers have returned the manuscripts, with their comments, the editor must then face the moment of truth.

Recently, I asked a distinguished historian of science to review a manuscript concerned with the history and philosophy of science. His review comprised only three sentences, yet it was one of the clearest and most useful reviews I have ever seen:

> Dear Bob:
>
> I had never before heard of [author's name] and from what there is in the book summary I really don't want to hear of him now. It seems to me very far removed from any idea I have of science, history, or, indeed, of philosophy. I wouldn't touch it with a barge pole.
>
> Cordially,

The Editor's Decision

Sometimes, the editor's decision is easy. If both reviewers advise "accept" with no or only slight revision, the editor has no problem. Unfortunately, there are many instances in which the opinions of the two reviewers are contradictory. In such cases, the editor either must make the final decision or send the manuscript out to one or more additional reviewers to determine whether a consensus can be established. The editor is likely to take the first approach if he or she is reasonably expert in the subject area of the manuscript and can thus serve as a third reviewer; the editor is especially likely to do this if the detailed commentary of one reviewer is considerably more impressive than that of the other. The second approach is obviously time-consuming and is used commonly only by weak editors; however, any editor must use this approach if the manuscript concerns a subject with which he or she is not familiar.

The review process being completed, and the editor having made a decision, on whatever basis, the author is now notified of the editor's decision. And it is the *editor's* decision. Editorial Board members and ad hoc reviewers can only recommend; the final decision is and must be the editor's. This is especially true for those journals (the majority) which use anonymous reviewers. The editor, assuming that he or she is of good character, will not hide behind anonymous reviewers. The decisions will be presented to the authors as though they were the editor's own, and indeed they are.

The editor's decision will be one of three general types, commonly expressed in one word as "accept," "reject," or "modify." Normally, one of these three decisions will be reached within 4 to 6 weeks after submission of the manuscript. If you are not advised of the editor's decision within 8 weeks, or provided with any explanation for the delay, do not be afraid to call or write the editor. You have the right to expect a decision, or at least a report, within a reasonable period of time; also, your inquiry may bring to light a problem. Obviously, the editor's decision could have been made but the missive bearing that decision could have been lost or delayed in the mail. If the delay was caused within the editor's office (usually by lack of response from one of the reviewers), your inquiry is likely to trigger an effort to resolve the problem, whatever it is.

Besides which, you should never be afraid to talk to editors. With rare exceptions, editors are awfully nice people. Never consider them adversaries. They are on *your* side. Their only goal as editors is to publish good science in understandable language. If that is not your goal also, you will indeed be dealing with a deadly adversary; however, if you share the same goal, you will find the editor to be a resolute ally. You are likely to receive advice and guidance that you could not possibly buy.

The Accept Letter

Finally, you get "the word." Suppose that the editor's letter announces that your manuscript has been accepted for publication. When you receive such a letter, you have every right to treat yourself to a double martini or whatever you do when you have cause both to celebrate and to admire yourself. The reason that such a celebration is appropriate is the relative rarity of the event. In the good journals (in biology at least), only about 5% of the manuscripts are accepted as submitted.

The Modify Letter

More likely, you will receive from the editor a bulky manila envelope containing both copies of your manuscript, two or more lists labeled "reviewers' comments," and a covering letter from the editor. The letter may say something like "Your manuscript has been reviewed, and it is being returned to you with the attached comments and suggestions. We believe these comments will help you improve your manuscript." This is the beginning phraseology of a typical modify letter.

By no means should you feel disconsolate when you receive such a letter. Realistically, you should not expect that rarest of all species, the accept letter without a request for modification. The vast majority of submitting authors will receive either a modify letter or a reject letter, so you should be pleased to receive the former rather than the latter.

When you receive a modify letter, examine it and the accompanying reviewers' comments carefully. (In all likelihood, the modify letter is a form letter, and it is the attached comments that are significant.) The big question now is whether you can, and are willing to, make the changes requested by the reviewers.

If both referees point to the same problem in a manuscript, almost certainly it *is* a problem. Occasionally, a referee may be biased, but hardly two of them simultaneously. If referees misunderstand, readers will. Thus, my advice is: If two referees misunderstand the manuscript, find out what is wrong and correct it before resubmitting the manuscript to the same journal or to another journal.

If the requested changes are relatively few and slight, you should go ahead and make them. As King Arthur used to say, "Don't get on your high horse unless you have a deep moat to cross."

If major revision is requested, however, you should step back and take a total look at your position. One of several circumstances is likely to exist.

First, the reviewers are right, and you now see that there are fundamental flaws in your paper. In that event, you should follow their directions and rewrite the manuscript accordingly.

Second, the reviewers have caught you off base on a point or two, but some of the criticism is invalid. In that event, you should rewrite the manuscript with two objectives in mind: Incorporate all of the suggested changes that you can reasonably accept, and try to beef up or clarify those points to which the reviewers (wrongly, in your opinion) took exception. Finally, and importantly, when you resubmit the revised manuscript provide a covering statement indicating your point-by-point disposition of the reviewers' comments.

Third, it is entirely possible that one or both reviewers and the editor seriously misread or misunderstood your manuscript, and you believe that their criticisms are almost totally erroneous. In that event, you have two alternatives. The first, and more feasible, is to submit the manuscript to another journal, hoping that your manuscript will be judged more fairly. If, however, you have strong reasons for wanting to publish that particular manuscript in that particular journal, do not back off; resubmit the manuscript. In this case, however, you should use all of the tact at your command. Not only must you give a point-by-point rebuttal of the reviewers' comments; you must do it in a way that is not antagonistic. Remember that the editor is trying hard, probably without pay, to reach a *scientific* decision. If you start your covering letter by saying that the reviewers, whom the editor obviously has selected, are "stupid" (I have seen such letters), Jimmy the Greek will give you 100 to 1 that your manuscript will be immediately returned without further consideration. On the other hand, *every* editor knows that *every* reviewer can be wrong and in time (Murphy's law) will be wrong. Therefore, if you dispassionately point out to the editor exactly why you are right and the reviewer is wrong (*never* say that the editor is wrong), the editor is very likely to accept your manuscript at that point or, at least, send it out to one or more additional reviewers for further consideration.

If you do decide to revise and resubmit the manuscript, try very hard to meet whatever deadline the editor establishes. Most editors do set deadlines. Obviously, many manuscripts returned for revision are not resubmitted to the same journal; hence, the journal's records can be cleared of dead wood by considering manuscripts to be withdrawn after the deadline date passes.

If you meet the editor's deadline, he or she may accept the manuscript forthwith. Or, if the modification has been substantial, the editor may return it to the same reviewers. If you have met, or defended your paper against, the previous criticism, your manuscript will probably be accepted.

On the other hand, if you fail to meet the deadline, your revised manuscript may be treated as a new manuscript and again subjected to full review, possibly by a different set of reviewers. Thus, it is wise to avoid this double jeopardy, plus additional review time, by carefully observing the editor's deadline if it is at all possible to do so.

The Reject Letter

Now let us suppose that you get a reject letter. (Almost all editors say "unacceptable" or "unacceptable in its present form"; seldom is the harsh word "reject" used.) Before you begin to weep, do two things. First, remind yourself that you have a lot of company; most of the good journals have reject rates approximating (or exceeding) 50%. Second, read the reject letter *carefully* because, like modify letters, there are different types of rejection.

Many editors would class rejections in one of three ways. First, there is (rarely) the total rejection, the type of manuscript that the editor "never wants to see again" (a phrase one forthright but not tactful editor put into a reject letter). Second, and much more common, there is the type of manuscript that contains some useful data but the data are seriously flawed. The editor would probably reconsider such a manuscript if it were considerably revised and resubmitted, but the editor does not recommend resubmission. Third, there is the type of manuscript which is basically acceptable, except for a defect in the experimental work—the lack of a control experiment perhaps—or for a major defect in the *manuscript* (the data being acceptable).

If your "rejection" is of the third type, you probably should consider the necessary repairs, as detailed in the reviewers' comments, and resubmit a revised version to the same journal. If you can add that control experiment, as requested by the editor, the new version will almost certainly be accepted. (Many editors reject a paper that requires additional experimentation, even though it might be easy to modify the

© 1982 United Feature Syndicate, Inc.

paper to acceptability.) Or, if you make the requested major change in the manuscript, e.g., totally rewriting the Discussion or converting a full paper to a note, your resubmitted manuscript is quite likely to be accepted.

If your rejection is of the second type (seriously flawed, according to the editor's reject letter and the reviewers' comments), you should probably not resubmit the same manuscript to the same journal, *unless* you can make a convincing case to the editor that the reviewers seriously misjudged your manuscript. You might, however, hold the manuscript until it can be buttressed with more extensive evidence and more clear-cut conclusions. Resubmission of such a "new" manuscript to the same journal would then be a reasonable option. Your covering letter should reference the previous manuscript and should state briefly the nature of the new material.

If your rejection is of the first (total) type, it would be pointless to resubmit the manuscript to the same journal or even to argue about it.

If the manuscript is really bad, you probably should not (re)submit it anywhere, for fear that publication might damage your reputation. If there is work in it that can be salvaged, incorporate those portions into a new manuscript and try again, but in a different journal.

Cheer up. You may some day have enough rejection letters to paper a wall with them. You may even begin to appreciate the delicate phrasing that is sometimes used. Could a letter such as the following possibly hurt? (This is reputedly a rejection slip from a Chinese economics journal.)

> We have read your manuscript with boundless delight. If we were to publish your paper, it would be impossible for us to publish any work of a lower standard. And as it is unthinkable that, in the next thousand years, we shall see its equal, we are, to our regret, compelled to return your divine composition, and to beg you a thousand times to overlook our short sight and timidity.

Editors as Gatekeepers

Perhaps the most important point to remember, whether dealing with a modify or a reject, is that the editor is a middleman between you and the reviewers. If you deal with the editor respectfully, and if you can defend your work scientifically, most of your "modifies" and even your rejects will in time become published papers. The editor and the reviewers are usually on your side. Their primary function is to help you express yourself effectively and provide you with an assessment of the science involved. It is to your advantage to cooperate with them in all ways possible.

Having spent the proverbial "more years than I care to remember" working with a great many editors, I am totally convinced that, were it not for the gatekeeper role so valiantly maintained by editors, our scientific journals would soon be reduced to unintelligible gibberish.

No matter how you are treated by editors, try somehow to maintain a bit of sympathy for members of that benighted profession. H. L. Mencken, one of my favorite authors (literary, that is), wrote a letter dated 25 January 1936 to William Saroyan, saying, "I note what you say about your aspiration to edit a magazine. I am sending you by this mail a six-chambered revolver. Load it and fire every one into your head. You will thank me after you get to Hell and learn from other editors how dreadful their job was on earth."

Guidelines for Reviewers

American Society for Microbiology

An unpublished manuscript is a privileged document. Please protect it from any form of exploitation. Do not cite a manuscript or refer to the work it describes before it has been published and do not use the information that it contains for the advancement of your own research or in discussions with colleagues.

Adopt a positive, impartial attitude toward the manuscript under review, with the aim of promoting effective and accurate scientific communication.

If you believe that you cannot judge a given article impartially, please return it immediately to the editor with an explanation.

Reviews must be completed by the date stipulated on the review form. If you know that you cannot finish the review within that time, immediately return the manuscript to the editor with a note of explanation. If possible, provide the names and addresses of two reviewers who are competent to handle the subject matter.

In your review, consider the following aspects of the manuscript:

Significance of research question or subject studied
Originality of work
Appropriateness of approach or experimental design
Adequacy of experimental techniques
Soundness of conclusions and interpretation
Relevance of discussion
Soundness of organization
Adherence to ASM style as set forth in Instructions to Authors
Adequacy of title and abstract
Appropriateness of figures and tables
Length of article
Adherence to correct nomenclature (genetic, enzyme, drug, bio-
chemical, etc.)
Appropriate literature citations

You are not requested to correct deficiencies of style or mistakes in grammar, but any help you can give in clarifying meaning will be appreciated. If you wish to mark the text of the manuscript, use a pencil or make a photocopy, mark it, and return it together with the original. If the editor decides to reject the paper, he will usually not want to return to the author a manuscript that has been heavily marked. Each accepted manuscript will be edited by the professional staff employed in the ASM Publications Office. It is their function to polish and correct the grammar, syntax, and spelling and to enforce the editorial style of the journal.

However, be on the lookout for errors which the copy editors (who are not microbiologists) might miss. Examples would be misspellings of chemical names, use of outmoded terminology, misspelled or misidentified scientific names of organisms, inappropriate scientific jargon, and incorrect genetic nomenclature.

You can be particularly helpful in pointing out unnecessary illustrations and data that are presented in *both* tabular (or graphic) form and in detail in the text. Such redundancies are a waste of space and readers' time and usually cannot be dealt with by the Publications Office staff.

A significant number of authors have not learned how to organize data and will benefit from guidance in this area. Editors often reject such manuscripts but encourage the author to rewrite the paper, to ask colleagues in the field for criticism, and after appropriate revision to resubmit it. Such simple, basic advice can be very valuable to inexperienced authors.

Do not discuss the paper with its authors. Although it may seem natural and reasonable to discuss points of difficulty or disagreement directly with the author, especially if you are generally in *favor* of publication and do not mind revealing your identity, this practice is prohibited because the other reviewer and the editor may have different opinions, and the author may be misled by having "cleared things up" with the reviewer who contacted him directly.

In your comments intended for transmission to the author, do not make any specific statement about the acceptability of a paper, but advise the editor of your recommendation by checking the review form at the appropriate place. *(The responsibility for acceptance or rejection lies with the editor.)* Suggested revisions should be stated as such and not expressed as conditions of acceptance. Present criticism dispassionately and avoid offensive remarks.

Organize your review so that an introductory paragraph summarizes the major findings of the article, gives your overall impression of the paper, and highlights the major short-comings. This paragraph should be followed by specific, numbered comments, which, if appropriate, may be subdivided into major and minor points. (The numbering facilitates the editor's evaluation of the author's rebuttal.)

Confidential remarks directed to the editor should be typed (or handwritten) on a separate sheet, not on the review form. You might want to distinguish between revisions considered essential and those judged merely desirable. Although you may advise the editor that you consider some revisions essential for acceptability, you should not so state to the author because the editor may not agree and/or the author may mistakenly conclude that adoption of the revision named will guarantee acceptance of the revised manuscript.

Your criticisms, arguments, and suggestions concerning the paper will be most useful to the editor and to the author if they are carefully documented. Do not make dogmatic, dismissive statements, particularly about the novelty of work. Substantiate your statements.

Reviewers' recommendations are gratefully received by the editor; however, since editorial decisions are usually based on evaluations derived from several sources, reviewers should not expect the editor to honor every recommendation.

Categories of recommendations: Accept, reject, modify, or convert to Note. Very few papers qualify for "accept" upon original submission. To be acceptable, a paper should be ready for publication except for minor style changes and perhaps corrections of grammar, spelling, etc., that can be taken care of by the Publications Office staff.

A paper that requires more than a trivial amount of experimental material (e.g., simple enzyme assay or growth rate determination) to complete the argument should be checked "reject."

When data and experiments appear sound and complete or nearly so but there are deficiencies in organization and presentation and interpretation of data, the choice of "modify" or "reject" depends on the extent of the problem. If the reviewer can follow the argument and feels that the author should be able, with reviewers' suggestions, to turn out a coherent revision within a month, the choice should be "modify." If the treatment is so muddled that the meaning of the data comes into question, or the material is incomprehensible to anyone but a specialist, the choice should be "reject."

Papers that present brief observations and much extraneous information should be converted to Note form. This does not mean, however, that figure legends and/or table footnotes should contain lengthy discussions of materials and methods. If these items cannot be handled in a short, concise manner (i.e., a few sentences), then a *short*, full-length paper would be the appropriate form.

Keep the pink copy of the review in your files. The manuscript may be returned to you for a second review, particularly when the requested modification was extensive or if the manuscript is resubmitted after rejection: you will need to evaluate the author's responses to your criticisms.

Chapter 17

The Publishing Process (How to Deal with Printers)

Proofread carefully to see
if you any words out.
—ANONYMOUS

Why Proof Is Sent to Authors

Some authors seem to forget their manuscripts as soon as they are accepted for publication, paying little attention to the galley proofs when they arrive and assuming that their papers will magically appear in the journals, without error.

Why is proof sent to authors? "Authors are provided with proof of their paper for one primary reason: to check the accuracy of the type-composition" ("Proofreading Instructions," *Journal of Bacteriology*). In other words, you should examine the proofs carefully for typographical errors. No matter how perfect your manuscript might be, it is only the printed version in the journal that counts. If the printed article contains serious errors, all kinds of later problems can develop, not the least of which may be irreparable damage to your reputation.

The damage can be real, in that many errors can totally destroy comprehension. Something as minor as a misplaced decimal point can sometimes make a published paper almost useless. In this world, we can be sure of only three things: death, taxes, and typographical errors.

Misspelled Words

Even if the error does not greatly affect comprehension, it won't do your reputation much good if it turns out to be funny. Readers will know what you mean if your paper refers to a "nosocomical infection," and they will get a laugh out of it, but *you* won't think it is funny.

While on the subject of misspellings, I recall the Professor of English who had the chance to make a seminal comment on this subject. A student had misspelled the word "burro" in a theme. In a marginal comment, the Professor wrote: "A 'burro' is an ass; a 'burrow' is a hole in the ground. One really should know the difference."

An advertiser, a major laboratory supply corporation, submitted an ad with a huge boldface headline which proclaimed that "Quality is consistant because we care." I certainly hope they cared more about the quality of their products than they did about the quality of their spelling.

Although all of us in publishing occasionally lose sleep worrying about typos, I take comfort in the realization that whatever slips by my eye is probably less grievous than some of the monumental errors committed by my publishing predecessors.

My all-time favorite typo occurred in a Bible published in England in 1631. The Seventh Commandment read: "Thou shalt commit adultery." I understand that Christianity became very popular indeed after publication of that edition. If that statement seems blasphemous, I need only refer you to another edition of the Bible, printed in 1653, in which appears the line: "Know ye that the unrighteous shall inherit the kingdom of God."

If you read proof in the same way and at the same speed that you ordinarily read scientific papers, you will probably miss 90% of the typographical errors.

The best way to read proof that I have found is, first, *read* it and, second, *study* it. The reading, as I mentioned, will miss 90% of the errors, but it will catch errors of *omission*. If the printer has dropped a line, reading for comprehension is the only likely way to catch it. Alternatively, or *additionally*, two people should read the proof, one reading aloud while the other follows the manuscript.

To catch most errors, however, you must slowly examine each word. If you let your eye jump from one group of words to the next, as it does in normal reading, you will not catch very many misspellings. Especially, you should study the technical terms. Remember that keyboard operators are not scientists. A good operator might be able to set the word "cherry" 100 times without error; however, I recall seeing a proof in which the word "*Escherichia*" was misspelled 21 consecutive times (in

96 *How to Write and Publish a Scientific Paper*

four different ways). I also recall wondering about the possible uses for a chemical whose formula was printed as $C_{12}H_6Q_3$.

I mentioned the havoc that could occur from a misplaced decimal point. This observation leads to a general rule in proofreading. Examine each and every number carefully. Be especially careful in proofing the tables. This rule is important for two reasons. First, errors frequently occur in keyboarding numbers, especially in tabular material. Second, you are the *only* person who can catch such errors. Most spelling errors are caught in the printer's proofroom or in the journal editorial office. However, these professional proofreaders catch errors by "eyeballing" the proofs; the proofreader has no way of knowing that a "16" should really be "61."

Marking the Corrections

When you find an error on a galley, it is important that the error be marked *twice*, once at the point where it occurs and once in the margin opposite where it occurs. It is the margin marks that the printer uses to identify the errors. A correction indicated only in the body of the typed material could easily be missed; the marginal notation is needed to call attention to it. This double marking system is illustrated in Fig. 9.

If you indicate your corrections clearly and intelligibly, the appropriate corrections will probably be made. However, you can reduce the chance of misunderstanding, and save time for yourself and all concerned, if you use established proofreaders' marks. These marks are a language universally used in all kinds of publishing. Thus, if you will take the time to learn just a few of the elements of this language, you will be able to use them in proofing any and all kinds of printing that you may be involved with throughout your career. The most common proofreaders' marks are listed in Table 10.

Additions to the Proofs

Early in this chapter, I stated that authors are sent proof so that they can check the accuracy of the typesetting. Stated negatively, "the proof stage is *not* the time for revision, rewriting, rephrasing, addition of more recent material, or any other significant change from the final edited manuscript" ("Proofreading Instructions," *Journal of Bacteriology*). There are three good reasons why you should not make substantial changes in the proofs.

First, an ethical consideration: Since neither proofs nor changes in the proof are seen by the editor unless the journal is a small one-person operation, it is simply not proper to make substantive changes. The

4 to picryl chloride of recipients sensitized with
5 picryl chloride, and cells from donors that had
6 been both *P. aeruginosa* injected and picryl
7 chloride sensitized failed to depress contact sen-
8 sitivity to oxazolone of recipient mice sensitized
9 with oxazolone. These results indicated that the
10 cells responsible for the depression of contact
11 sensitivity in *P. aeruginosa-injected* mice were
12 antigen specific in that they required specific
13 antigenic stimulation.

14 **Effect of cyclophosphamide on the pre-**
15 **cursors of suppressor cells in *P. aerugi-**
16 ***nosa*-injected mice.** Normal mice were sensi-
17 tized with oxazolone and 1 h later were injected
18 intravenously with 50×10^6 spleen cells from
19 donors sensitized 4 days previously with the
20 same antigen. Two groups of donors were also
21 injected with either *P. aeruginosa* or 200 mg of
22 cyclo./phosphamide per kg 24 or 48 h before
23 sensitization respectively. A third group of do-
24 nors received both *P. aeruginosa* and cyclophos-
25 phamide. Sensitized mice receiving no cells were
26 used as controls. The challenge of the experi-
27 mental and control groups was performed with
28 oxazolone 6 days after the cell transfer. Cyclo
29 phosphamide completely inhibited the develop-
30 ment of suppressor activity in the spleens of
31 mice injected with P. aeruginosa and sensitized
32 with oxazolone (Table 3).

33

DISCUSSION

34 The results show that heat killed *P. aerugi-*
35 *nosa* depresses contact sensitivity to oxazolone
36 in C57BL/6 mice when injected intravenously 24
37 h before sensitization. The spleens and the
38 draining lymph nodes of mice exhibiting an im-
39 paired reactivity to oxazolone contain a cell pop-
40 ulation capable of passively transferring the sup-
41 pression of contact sensitivity to recipients sen-
42 sitized immediately before the cell transfer with
43 the same antigen. The suppressive activity of
44 these cells appears to be antigen specific, since
45 they do not effect the response to a different
46 sensitizing agent, picryl chloride, and because
47 they arise in *P. aeruginosa*-injected mice only
48 when they are sensitized. These suppressor cells,
49 which occur in the draining lymph nodes and
50 spleen at 3 and 4 days after sensitization, respec-
51 tively, have precursors sensitive to cyclophos-
52 phamide.

FIG. 9. *A corrected galley proof. (Appreciation is expressed to Waverly Press, Inc., for typesetting this defective sample. A normal galley from Waverly would have so few errors that it would be useless for illustrative purposes.)*

97

TABLE 10. *Frequently used proofreaders' marks*

Instruction	Mark in text	Mark in margin
Capitalize	Hela cells	*cap*
Make lower case	the Penicillin reaction	*l.c.*
Delete	a very good reaction	*e*
Close up	Mac Donald reaction	⌒
Insert space	lymph node cells	#
Start new paragraph	in the cells¶The next	¶
Insert comma	in the cells after which	⋀
Insert semicolon	in the cells however	⋀;
Insert hyphen	well known event	=
Insert period	in the cells Then	⊙
Insert word	in cells	# *the* #
Transpose	proofreader	*tr*
Subscript	CO2	⋀2
Superscript	32P	⌄32
Set in roman type	The *bacterium* was	*rom*
Set in italic type	P. aeruginosa cells	*ital*
Set in boldface type	Results	*b.f.*
Let it stand	a very good reaction	*stet*

manuscript approved by the editor, after peer review, is the one that should be printed, not some new version containing material not seen by the editor and the reviewers.

Second, it is not wise to disturb typeset material, unless it is really necessary, because new typographical errors may be introduced. If a word is added to a line, many following lines may have to be reset also (to maintain even or "justified" margins). Incidentally, if you must add a word, look for a word of about the same size, in the same line or on the line above or below the addition, that you could delete to make room for the added word; or, if you need to delete a word, try to add a filler word to take up the space.

Third, corrections are expensive (15, 17). Because they are expensive, you should not abuse the publisher (possibly a scientific society of which you are an otherwise loyal member); in addition, you just might be hit with a substantial bill for author's alterations. Most journals absorb the cost of a reasonable number of author's alterations, but many, especially those with managing editors or business managers, will sooner or later crack down on you if you are patently guilty of excessive alteration of the galleys.

One type of addition to the proof is frequently allowed. The need arises when a paper on the same or a related subject appears in print while yours is in process. In light of the new study, you might be tempted to rewrite several portions of your paper. You must resist this temptation, for the reasons stated above. What you should do is prepare a short (a few sentences only) Addendum in Proof, describing the general nature of the new work and giving the literature reference. The Addendum can then be printed at the end without disturbing the body of the paper.

Addition of References

Quite commonly, a new paper appears that you would like to add to your Literature Cited, but you need not make any appreciable change in the text, other than adding a few words, perhaps, and the number of the new reference. (This assumes that the journal employs the numbered, alphabetized list system.)

Now hear this. If you add a reference at proof, do *not* renumber the references in your Literature Cited. Many, if not most, authors make this mistake, and it is a serious mistake. It is a mistake because the many changes then necessary in the reference list and in the text, wherever the cited numbers appear, involve significant printing cost; new errors may be introduced when the affected lines are rekeyboarded; and, almost certainly, you will miss at least one of the text references. The old number(s) will then appear in print, adding confusion to the literature.

What you *should* do is add the new reference with an "*a*" number. If the new reference would alphabetically fall between references 16 and 17, enter the new reference as "16*a*." In that way, the numbering of the rest of the list need not be changed.

Proofing the Illustrations

It is especially important that you examine carefully the proofs of the illustrations. Normally, the original manuscript *and the original illustrations* are returned to the author along with the proof. Thus, although you can depend on the proofreaders in the journal editorial office to backstop you in looking for typographical errors, *you* must decide whether the illustrations have been reproduced effectively because you have the originals with which the proofs must be compared.

If your paper contains important fine-structure photographs, and if you chose that particular journal because of its reputation for high-quality reproduction standards (fine screens, coated stock), you should not only expect almost faultless fidelity, you should also demand it. And you are the only one who can, because you are the one with the originals. You and you alone must serve as the quality control inspector.

Seldom will there be a problem with graphs and other line drawings, unless the copy editor has sized them so small that they are illegible or, rarely, misfigured the percentage reduction on one of a related group, so that it does not match.

With photographs, however, there are problems on occasion and it is up to *you* to spot them. Compare the illustration proof with the original. If the proof is darker overall, it is probably a simple matter of overexposure; if detail has thereby been lost, you should of course ask the printer to reshoot the photograph. (Don't forget to return the original illustration along with the proof.)

If the proof is lighter than the copy, it has probably been underexposed. It may be, however, that the "printer" (I use the word "printer" as shorthand for all of the many occupations that are involved in the printing process) purposely underexposed that shot. Sometimes, especially with photographs having very little contrast, underexposure will retain more fine detail than will normal exposure. Thus, your comparison should not really be concerned with exposure level but with fidelity of detail.

It may be that one area of the photograph is of particular importance. If that is so, and if you are unhappy with the reproduction, tell the printer, via marginal notes or by use of an overlay, exactly which part of the proof is lacking detail that is evident on the photograph. Then the printer will be able to focus on what is important to you.

When to Complain

If you have learned nothing else from this chapter, I trust that you now know that *you* must provide the quality control in the printing of illustrations in journals. In my experience, too many authors complain after the fact (after publication) without ever realizing that only they could have prevented whatever it is they are complaining about. For example, authors many times complain that their pictures have been printed upside-down or sideways. When I have checked out such complaints, I have found in almost all instances that the part of the photograph marked "top" on the proof was also the top in the journal; the author simply missed it. Actually, the author probably missed twice, once by neglecting to mark top on the photograph submitted to the journal and again by failing to note that the printer had marked top on the wrong side of the proof.

So, if you are going to complain, do it at the proof stage. And, believe it or not, your complaint is likely to be received graciously. Those of us who pay the bills realize that we have invested heavily in setting the specifications that can provide quality reproduction. We need your quality control, however, to insure that our money is not wasted.

The good journals are printed by good printers, hired by good publishers. The published paper will have your name on it, but the reputations of both the publisher and the printer are also at stake. They expect you to work with them in producing a superior product.

Because managing editors of such journals must protect the integrity of the product, those I have known would *never* hire a printer exclusively on the basis of low bids. John Ruskin was no doubt right when he said, "There is hardly anything in the world that somebody cannot make a little worse and sell a little cheaper, and the people who consider price only are this person's lawful prey."

A sign in a job printing shop made the same point:

PRICE
QUALITY
SERVICE
(pick any two of the above)

Chapter 18

The Electronic Manuscript

Be not the first by whom the new are tried,
Nor yet the last to lay the old aside.
— ALEXANDER POPE

Capturing Keystrokes

For many years, scientific papers were typed on a typewriter in the author's institution and then "retyped" by the printer. This second typing, known as "type composition," was for some generations done by lead-casting machines such as the Linotype and Monotype. These "hot lead" machines have been replaced in the past 20 years by photocomposition devices. With photocomposers, a photographic image of a keyboarded letter is produced rather than a metal casting. Normally, however, the "system" has remained the same in that the printer re-keyboards (retypes) the entire manuscript, which is a labor-intensive and expensive operation. Typically, about half of the total manufacturing cost of a specialized (i.e., low circulation) scientific journal is the cost of composition. This traditional and almost universally used system of re-keyboarding the author's manuscript has been described in Chapter 14 ("How to Type the Manuscript") and Chapter 17 ("The Publishing Process"). I truly believe, however, that this system will die, and I believe that death is imminent.

The main parts of the new system are already in place. At the printer's end, we have highly sophisticated computer-driven photocomposition devices. At the author's end, a great number (perhaps now a majority) of institutions have electronic devices (word-processing units) whose output can serve as input to the printer's photocomposers. Further, with the millions of relatively inexpensive personal computers now flooding the market, scientists like other professionals will find many uses for them. For data storage, for online literature searching, and for manuscript preparation, the new microcomputers are ideal for scientific uses. And, as these devices become more "user-friendly," scientists will not hesitate to use them. Just as many scientists use Xerox machines every day without really knowing how they work, so they will soon be using computers in the same way.

What has been missing is the ability for machines to talk to each other. With printers having a variety of photocomposition devices in use, and with upwards of 100 different word-processing models on the market, we seem to be living in a polyglot boarding house. Although the output of almost any word processer can be coded to drive any photocomposer, the expense may well be higher than re-keyboarding. What is needed is a universal language that all machines can understand.

Such a language is now under development. The Association of American Publishers proposes to develop industry standards and author guidelines for the preparation of electronic manuscripts. As I understand it, a main goal of the AAP study is to develop "generic coding" for manuscripts. When such generic coding is established, almost all printers will be able to receive the output (magnetic tape, floppy disk) of the author's word-processing unit and use it as direct input to the photocomposer. Thus, by capturing the author's keystrokes, the considerable expense of the second keyboarding previously supplied by the printer (but paid for by the publisher) can be avoided. The next generation of authors (and many of this one) will be inputting their manuscripts into word-processing systems, in their homes or in their offices, and typeset galleys will result therefrom.

Working at a Terminal

Many scientists are already using an electronic terminal, personal computer, or stand-alone word processer. Although relatively few scientists can as yet provide output that can be used as direct manuscript input by the printer (because of the coding problem), there are other advantages in using a terminal. Keyboarding at a terminal is scarcely more difficult than keyboarding at a standard typewriter, and correction and manipulation of the keyboarded material makes the terminal (with video display) vastly superior to the typewriter, since the editing function provides much additional power.

Most important, the terminal has a memory. While writing at the terminal, you can conveniently call up anything you have already stored.

You can refer to the raw data in your lab notebooks, to relevant papers or abstracts, and to reviewer's or editor's comments you may have received.

You can also use your terminal to tap into the vast memory of other computers. There are now dozens of data bases that scientists can use for various purposes, notably for searching the literature. Direct online searching of *Index Medicus* (National Library of Medicine), *Science Citation Index* (Institute for Scientific Information), *Biological Abstracts*, and *Chemical Abstracts* can now readily be done (for a fee, of course) from home or office.

Writing at a Terminal

The terminal keyboard looks like a typewriter keyboard and it works like a typewriter keyboard. If you can use a typewriter (even if you are of the hunt and peck school), you can use a terminal. However, you can do much more at a terminal. By learning how to use the appropriate function keys, you can add, delete, and reposition words, sentences, paragraphs, and entire sections. When you make any change, you can immediately examine the display screen for the accuracy of the correction. You need check only the corrected portion of the manuscript; it is not necessary to proofread the entire new version of the manuscript. By using a terminal, you can go through a number of revisions of the manuscript at one sitting if desired. And, at any time, the attached printer can be used to produce a "hard copy" (a "typewritten" copy on paper). Thus, the long delays in getting new versions of the manuscript typed, and proofread, can be avoided.

Display terminals can also be used to generate various kinds of graphs and charts. Still another function is "electronic mail," whereby scientists can use their terminals to communicate with other scientists in an established network. It would be easy, for example, for co-authors of a paper to exchange sections of a manuscript, questions, comments, references, and the like.

At the end of the writing process, when the author(s) has decided that the manuscript is ready for submission to a journal, a "button" is pushed and a hard copy of the final manuscript is produced, ready for mailing. Then, during the review process, corrections can be made as required to the manuscript stored in the microcomputer memory. Finally, the edited, accepted version of the manuscript, in electronic form, can be sent to the printer. If the printer can use this electronic manuscript as direct input to the photocomposer, as will soon be the norm, typeset galleys can be produced rapidly, inexpensively, and (theoretically) without error.

Chapter 19

How to Order and Use Reprints

I shot a poem into the air,
It was reprinted everywhere
From Bangor to the Rocky Range
And always credited to—Exchange.
    —FRANKLIN PIERCE ADAMS

How to Order Reprints

Reprints are to some degree a vanity item. "Vanity of vanities; all is vanity" (Ecclesiastes I, 2; XII, 8). Having said that, I will now give a few words of advice on how to buy reprints and use them, because I know you will. Everybody does. Although Garfield (19) and others have described the reprint system as obsolete, it is a rare author indeed who does not order reprints. And the rare exception is usually someone from a newly developing (i.e., poor) country who simply cannot afford them.

The "how to order" is usually easy. A reprint order blank is customarily sent with the galley proofs. In fact, this custom is so universal that you should call or write the journal office if the reprint order form does *not* arrive with the proof, because the omission was probably inadvertent.

Reprint Manufacturing Systems

Some journals still manufacture reprints by the "run with journal" process. (The reprints are printed as an overrun while the journal itself is being printed.) In that process, it is important that you get your order in early. Return the order form, with the galleys if so directed, at an early time, rather than waiting for an official purchase order to grind through your institution's mills. Try to get a purchase order *number* even though there might be delay in getting the purchase order itself.

Some journal reprints are now manufactured on small offset presses, in a process essentially unrelated to manufacture of the journal. In recent years, the cost of paper has increased tremendously; the wastage of paper inherent in the "run with journal" system has made that system economically unsound.

The newer system has one huge advantage: Reprints of your paper can be produced at any time in any number. Therefore, if you publish in such a journal, you need never worry about running out of reprints.

How Many to Order

Even though you may be able to reorder later, it is wise to overorder in the first instance. Most journals charge a substantial price for the first 100 reprints, to cover the setup and processing costs. The second 100 is usually very much cheaper, the modest increase in price reflecting only the cost of additional paper and press time. Therefore, if you think you may need about 100 reprints, order 200; if you might need 200, order 300. The price differential is so slight that it would be foolish not to err on the high side. The price list shown in Table 11 is typical of many.

How to Use Reprints

As for using reprints, you may let your imagination and vanity be your guide. Start by sending one to your mother because that is easier than writing the letter that you should have written long ago. If it is really a good paper, send a reprint to anybody you want to impress, especially any senior colleagues who may some day be in a position to put in a good word for you.

TABLE 11. *Reprint price list: Journal of Bacteriology*

Pages	Number of copies					
	100	200	300	400	500	Additional 100's
4	$116	$132	$146	$160	$173	$12
8	178	199	217	234	252	15
12	249	275	300	323	345	20
16	331	362	392	422	450	26
20	414	451	487	521	554	31

Your main consideration is whether or not to play the "postcard" game. The game can be simplified and made more efficient by using Request-A-Print (RAP) Cards, available from the Institute for Scientific Information. Some scientists refuse to play the game, using instead a distribution list which they believe will get the reprints to colleagues who might really need the reprints. Routine postcards or form letters are ignored, although almost everyone would respond favorably to a personal letter.

Yet, although many scientists resent the time and expense of playing the postcard game, most of them play it anyway. And, vanity aside, the game may occasionally be worth the reprint. If so, the reasons may be somewhat as follows.

The largest number of reprint requests will come from people who can best be defined as "collectors." They tend to be "library" scientists, possibly graduate students or postdoctoral fellows, who are likely to have a wide interest in the literature and perhaps considerably less interest in laboratory manipulation. You probably won't recognize the names, even if you can read the signatures at the bottoms of the cards, because these individuals probably have not published in your field (if they have published at all). In time, you may begin to recognize some of the names, because the real collector collects with dogged determination. Every time you publish, you are likely to receive reprint requests from the same band of collectors working your particular subject area.

If you can recognize the collector, should you respond? Probably. There is, I think, room in science for the multidisciplinary types who spend hours in the library, constantly collecting, organizing, and synthesizing broad areas of the literature. Such broad-based people may not be at the forefront of research science, but often they become good teachers or good administrators; and, in the meantime, they are very likely to produce one or more superb review papers or monographs, often on a cosmic subject that only a collector would know how to tackle.

The next largest group of reprint requests is likely to come from foreign countries or from very small institutions. Quite obviously, these people have seen your paper listed in one of the indexing or abstracting services, but have not seen the paper itself because the journal is not available within their institution. (Expect a surge in requests within days after your paper is listed in *Current Contents*.) Should you respond to such requests? Frankly, if you send out reprints at all, I think that this group merits first consideration.

The third group of requests will come from your peers, people you know or names or laboratories that you recognize as being involved in your own or a related field. Should you respond to such requests? Probably, because you know that the reprints will actually be used. Your main concern here is whether it might be better to prepare a mailing list, so that you and some of your colleagues can exchange reprints without wasting time and expense with the requests.

Should *you* collect reprints? If so, how? That, of course, is up to you, but a few guidelines may be helpful.

You should realize, at the outset, that reprints are useful, if at all, as a *convenience*. Unlike books and journals, they have absolutely no economic value. I have known of several prominent scientists who, upon retirement, were upset because their vast reprint collections could not be sold, no institution would accept them as a gift, and even scrap paper dealers refused them because of the staples.

How to File Reprints

So, if reprints are to be used for your personal convenience, what would be convenient? Consider arranging your reprints alphabetically by *author* (cross-indexing additional authors). Most scientists seem to prefer a subject arrangement, but, as the collection grows, as subjects and interests change, and as time passes, more and more of the collection becomes inaccessible. As a former librarian, I assure you that every subject system ever devised will break down in time, and I also assure you that there is nothing so maddening as to search fruitlessly for something that you need and that you *know* you have somewhere.

If you have or expect to have a large reprint collection, no simple filing system will provide efficient retrievability. Records must be established. The records (probably on 3 × 5 cards) can be kept in a number of ways. Cards may be established in brief form for authors and co-authors and for any number of subject entries. All cards are maintained in one dictionary catalog (shoebox?). The reprints themselves are most likely filed by accession number, with that number being recorded on all relevant author and subject cards. Such record keeping is relatively easy and surprisingly efficient.

If you have a microcomputer, you can have the ultimate in retrievability. With a microcomputer and the Sci-Mate Personal Data Manager software (available from the Institute for Scientific Information), you can search by any characters, word, name, or phrase that might be anywhere in any entry. All previous manual filing systems pale in comparison.

What to Collect

What reprints should you collect? Let us get to the heart of the matter, or at least the aorta. Unless you are really a collector by personality, you should limit your collection to those items that are *convenient*. Knowing that you cannot collect everything, the best rule is to collect the difficult. You should not collect reprints of papers published in journals that you own, and you probably should not collect reprints from journals that are readily available in almost all libraries. You *should* collect reprints of papers published in the small, especially foreign, journals or in conference proceedings or other offbeat publications. Thus, measured in terms of convenience, your reprint collection need not supplant the library down the hall, but it is a convenience to have access in your own files to material that is not available in the library. Besides, the reprints are *yours*; you can mark them up, cut them up, and file them in any way that you find useful.

Your reprint file may also be used to house the photocopies of journal articles that you obtain. If your library obtains for you a photocopy of an article, via an interlibrary transaction, obviously that is exactly the kind of item that should go in your collection (because it would be convenient not to have to go through the interlibrary loan process again). Or, you may obtain copies of articles listed in *Current Contents* via ISI's OATS (original article text service); if you do, these would also be candidates for your collection (because it would be convenient to escape the trouble and the fee the next time you need that article).

Chapter 20

How to Write a Review Paper

Nature fits all her children with something to do,
He who would write and can't write, can surely review.
—JAMES RUSSELL LOWELL

Characteristics of a Review Paper

A "review" paper is *not* an original publication. On occasion, a review will contain new data (from the author's own laboratory) that have not yet appeared in a primary journal. However, the purpose of a review paper is to review previously published literature and to put it into some kind of perspective.

A review paper is usually long, typically ranging between 10 and 50 printed pages. The subject is fairly general, compared with that of research papers. And the literature review is, of course, the principal product. However, the really good review papers are much more than annotated bibliographies. They offer critical evaluation of the published literature and often provide important conclusions based on that literature.

The organization of a review paper is different from that of a research paper. Obviously, the Materials and Methods, Results, Discussion arrangement cannot be used for the review paper.

If you have previously written research papers and are now about to write your first review, it might help you conceptually if you visualize the review paper as a research paper, as follows. Greatly expand the

Introduction; delete the Materials and Methods (unless original data are being presented); delete the Results; and expand the Discussion.

Actually, you have already written many review papers. In format, a review paper is not very different from a well-organized term paper or thesis.

Like a research paper, however, it is the *organization* of the review paper that is important. The writing will almost take care of itself if you can get the thing organized.

Preparing an Outline

Unlike research papers, there is no prescribed organization for review papers. Therefore, you will have to develop your own. The cardinal rule for writing a review paper is *prepare an outline.*

The outline must be prepared carefully. The outline will assist you in organizing your paper, which is all-important. If your review is organized properly, the overall scope of the review will be well defined and the integral parts will fit together in logical order.

Obviously, you must prepare the outline *before* you start writing. Moreover, *before* you start writing, it is wise to determine whether a review journal would be interested in such a manuscript. Possibly, the editor will want to limit or expand the scope of your proposed review or to add or delete certain of the subtopics.

The "Instructions to Authors" in *Microbiological Reviews* says it this way: "Prospective authors are advised to discuss with the Editor the suitability of their proposed contribution. The preparation of a synopsis or topical outline is desirable, since it often elicits constructive suggestions from editorial consultants. In addition, a list of key references showing the author's contributions to the field and a one- or two-paragraph statement detailing the aim, scope, and relevance of the topic to be reviewed must be included with the outline."

Not only is the outline essential for the preparer of the review, it is also very useful to potential readers of the review. For that reason, many review journals print the outline at the beginning of the article, where it serves as a convenient table of contents for prospective readers. A well-constructed outline is shown in Fig. 10.

Types of Reviews

Before actually writing a review, you also need to determine the critical requirements of the journal to which you plan to submit the manuscript. That is because some journals demand critical evaluation of the literature, whereas others are more concerned with bibliographic

MICROBIOLOGICAL REVIEWS, Sept. 1978, p. 592-613
0146-0749/78/0042-0592$02.00/0
Vol. 42, No. 3
Printed in U.S.A.

Pathophysiological Effects of *Vibrio cholerae* and Enterotoxigenic *Escherichia coli* and Their Exotoxins on Eucaryotic Cells

KAREN L. RICHARDS AND STEVEN D. DOUGLAS*

Departments of Microbiology and Medicine, University of Minnesota Medical School, Minneapolis, Minnesota 55455

FIG. 10. *Outline of a review paper.*

completeness. There are also matters of organization, style, and emphasis that you should have in mind before you proceed very far.

By and large, the old-line review journals prefer and some demand authoritative and critical evaluations of the published literature on a subject. Many of the "book" series ("Annual Review of," "Recent Advances in," "Yearbook of," etc.), however, publish reviews designed to compile and to annotate but not necessarily to evaluate the papers published on a particular subject during a defined time period. Some active areas of research are reviewed on a yearly basis. Both of these types of review papers serve a purpose, but the different purposes need to be recognized.

At one time, review papers tended to present historical analyses. In fact, the reviews were often organized in chronological order. Although this type of review is now less common, one should not deduce that the history of science has become less important. There is still a place for history.

Today, however, most review media prefer either "state of the art" reviews or reviews that provide a new understanding of a rapidly moving field. Only the recent literature on the subject is catalogued or evaluated. If you are reviewing a subject that has not previously been reviewed or one in which misunderstandings or polemics have developed, a bit more coverage of the historical foundations would be appropriate. If the subject has been effectively reviewed before, the starting point for your review might well be the date of the previous review (not publication date, but the actual date up to which the literature has been reviewed).

Writing for the Audience

Another basic difference between review papers and primary papers is the *audience*. The primary paper is highly specialized and so is its audience (peers of the author). The review paper will probably cover a number of these highly specialized subjects, so that the review will be read by many peers. The review paper will also be read by many people in related fields, because the reading of good reviews is the best way to keep up in one's broad areas of interest. Finally, review papers are valuable in the teaching process, so that student use is likely to be high. (For these reasons, by the way, order *plenty* of reprints of any review paper you publish, because you will be inundated with reprint requests.)

Because the review paper is likely to have a wide and varied audience, your style of writing should be much more general than it need be for a research paper. Jargon and specialized abbreviations must be eliminated or carefully explained. Your writing style should be expansive rather than telegraphic.

Importance of Introductory Paragraphs

Readers are much influenced by the Introduction of a review paper. They are likely to decide whether or not to read further on the basis of what they find in the first few paragraphs (if they haven't already been repelled by the title).

Readers are also influenced by the first paragraph of each major section of a review, deciding whether to read, skim, or skip the rest of the section depending on what they find in the first paragraph. If "first paragraphs" are well written, all readers, including the skimmers and skippers, will be able to achieve some degree of comprehension of the subject.

Importance of Conclusions

Because the review paper covers a wide subject for a wide audience, a form of "conclusions" is a good component to consider and take the trouble to write. This is especially important for a highly technical, advanced, or obscure subject. Painful compromises must sometimes be made, if one really tries to summarize a difficult subject to the satisfaction of both expert and tyro. Yet, good summaries and simplifications will in time find their way into textbooks and mean a great deal to students yet to come.

Award for Excellence

The importance of review papers has recently been recognized by the National Academy of Sciences in establishing the James Murray Luck Award for Excellence in Scientific Reviewing. Co-sponsored by the Institute for Scientific Information and Annual Reviews, Inc., the Award, consisting of a scroll and $5,000, is given annually to an author of a particularly meritorious scientific review.

Chapter 21

How to Write a Conference Report

*Reading maketh a full man, conference
a ready man, and writing an exact man.*
—FRANCIS BACON

Definition as Nonprimary Publication

A conference report can be of many kinds. However, let us make a few assumptions and, from these, try to devise a picture of what a more-or-less typical conference report should look like.

It all starts, of course, when you are invited to participate in a conference (congress, symposium, workshop, panel discussion, seminar, colloquium), the proceedings of which will be published. At that early time, you should stop to ask yourself, and the conference convener or editor, exactly what is involved with the publication.

The biggest question, yet one that is often left cloudy, is whether the proceedings volume will be defined as primary. If you or other participants present previously unpublished data, the question arises (or at least it should) as to whether data published in the proceedings have been validly published, thus precluding later republication in a primary journal.

As more and more scientists, and their societies, become aware of the need to define their publications, there will be fewer problems. For one thing, conferences have become so popular in recent years that the conference report literature has become a very substantial portion of the total literature in many areas of science.

The clear trend, I think, is to define conference reports as not validly published primary data. This is seemingly in recognition of three important considerations: (i) Most conference proceedings are one-shot, ephemeral publications, not purchased widely by science libraries around the world; thus, because of limited circulation and availability, they fail one of the fundamental tests of valid publication. (ii) Most conference reports are essentially review papers, which do not qualify as primary publications, or they are preliminary reports, presenting data and concepts that may still be tentative or inconclusive and which the scientist would not yet dare to contribute to a primary publication. (iii) Conference reports are normally not subjected to peer review or to more than minimal editing; therefore, because of lack of any real quality control, many reputable publishers now define proceedings volumes as nonprimary.

This is important to you, so that you can determine whether or not your data will be buried in an obscure proceedings volume. It also answers in large measure how you should write the report. If the proceedings is adjudged to be primary, you should (and the editor will no doubt so indicate) prepare your manuscript in journal style. You should give full experimental detail, and you should present both your data and your discussion of the data as circumspectly as you would in a prestigious journal.

If, on the other hand, you are contributing to a proceedings which is not a primary publication, your style of writing may be (and should be) quite different. The fundamental requirement of reproducibility, inherent in a primary publication, may now be ignored. You need not, and probably should not, have a Materials and Methods section. Certainly, you need not provide the intricate detail that might be required for a peer to reproduce the experiments.

Nor is it necessary to provide the usual literature review. Your later journal article will carefully fit your results into the preexisting fabric of science; your conference report should be designed to give the news and the speculation for today's audience. Only the primary journal need serve as the official repository.

Extended Abstract Concept

If your conference report is not a primary scientific paper, just how should it differ from the usual scientific paper? Overall, it should be of a size that is increasingly specified for a "synoptic" or extended abstract.

I prefer the term "extended abstract" because its meaning is readily apparent to authors, and also because the *American National Standard for Synoptics* (5) defines "synoptic" as "first publication . . . in a primary journal. . . ."

An extended abstract is usually limited to one or two printed pages, or 1,000 to 2,000 words. Usually, authors can be provided with a simple formula, such as: "up to five manuscript pages, double-spaced, and not more than three illustrations (any combination of tables, graphs, or photographs)." It was that exact formula that was provided to participants of the 10th International Congress of Chemotherapy (Zurich, 1977). The resultant Proceedings (35) comprised extended abstracts that averaged almost exactly two printed pages. The Proceedings appeared (within five months of the Congress) in two volumes containing close to 700 papers on 1,350 pages, at a list price of $75. By comparison, the Proceedings of the preceding Congress (London, 1975) appeared (a year after the Congress) as full papers, in eight massive volumes, at a crushing price of $256. Librarians are beginning to question purchase of high-priced proceedings, especially those that have been delayed one or more years in publication.

Presenting the New Ideas

As stated above, the conference report can be relatively short because most of the experimental detail and much of the literature review can be eliminated. In addition, the results can usually be presented in brief form. Because the full results will presumably be published later in a primary journal, only the highlights need be presented in the conference report.

On the other hand, the conference report might give greater space to *speculation*. Editors of primary journals can get quite nervous about discussion of theories and possibilities that are not thoroughly buttressed by the data. The conference report, however, should serve the purpose of the true preliminary report; it should present and encourage speculation, alternative theories, and suggestions for future research.

Conferences themselves can be exciting precisely because they do serve as the forum for presentation of the very newest ideas. If the ideas are truly new, they are not yet fully tested. They may not hold water. Therefore, the typical scientific conference should be designed as a sounding board, and the published proceedings should reflect that ambience. The strict controls of stern editors and peer review are fine for the primary journal but are out of place in the conference literature.

The typical conference report, therefore, need not follow the usual

Introduction, Materials and Methods, Results, Discussion progression that is standard for the primary research paper. Instead, the abstract approach is extended. The problem is stated; the methodology used is stated (but not described in detail); and the results are presented briefly, in one, two, or three tables or figures. Then, the meaning of the results is speculated about, often at considerable length. The literature review most likely involves description of related or planned experiments in the author's own laboratory or in the laboratories of colleagues who are currently working on related problems.

Editing and Publishing

Finally, it is only necessary to remind you that the editor of the proceedings, usually the convener of the conference, is the sole arbiter of questions relating to manuscript preparation. If the editor has distributed Instructions to Authors, you should follow them (assuming that you want to be invited to other conferences). You may not have to worry about rejection, since conference reports are seldom rejected; however, if you have agreed to participate in a conference, you should then follow whatever rules are established. If all contributors follow the rules, whatever they are, the resultant volume is likely to exhibit reasonable internal consistency and be a credit to all concerned.

If you are involved with the editing or publishing of a proceedings volume, you should make sure that a published copy is sent to the Institute for Scientific Information for inclusion in the *Index to Scientific and Technical Proceedings*.

Chapter 22

How to Write a Book Review

Books must follow sciences,
and not sciences books.
> —Francis Bacon

Scientific Books

Books are important in all professions, but they are especially important in the sciences. That is because the basic unit of scientific communication, the primary research paper, is short (typically five to eight printed pages in most fields) and narrowly specific. Therefore, to provide a general overview of a significant slice of science, writers of scientific books organize and synthesize the reported knowledge in a field into a much larger, more meaningful, package. In other words, new scientific knowledge is made meaningful by sorting and sifting the bits and pieces to provide a larger picture. Thus, the individual plants and flowers, and even the weeds, become a landscape.

Scientific, technical, and medical (STM) books are of many types. In broad categories, they can be considered as monographs, reference books, textbooks, and trade books. Because there are significant differences among these four types, a reviewer should understand the distinctions.

Monographs. Monographs are the books most used by scientists. Monographs are written by scientists for scientists. They are specialized and detailed. In form, they are often the equivalent of a long review article, and in fact some are labeled as such (annual reviews of, advances in, etc.). Some monographs are written by single authors; most are written by multiple authors. If a large number of authors are contributing to a monograph, there will be one or more editors who assign the individual topics and then edit the contributions to form a well-integrated volume. (This is the theory but not always the practice.) Such a monograph can be put together "by mail"; alternatively, a "conference" is called, papers are read, and a resultant volume contains the "proceedings."

As a publisher of long if not good standing, I now express a pet peeve. If, as a book reviewer, you want to comment about "the outrageously high price charged by the publisher," know what you are talking about or shut up. (That, by the way, is a good general rule for all aspects of book reviewing.) My point is this: Some reviewers have a simplistic notion about book prices; some even use a simplistic formula, saying perhaps that any book priced at less than 10 cents a printed page is O.K. but that a price higher than 10 cents a page "is gouging the scientific community." The fact of the matter is that the prices of books do and must vary widely; the variance depends primarily not on the size of the book but on the size of the audience. A book with potential sales of 10,000 or more copies can be priced modestly; a book with potential sales of 1,000 to 2,000 copies *must* carry a high price, if the publisher is to stay in business. Thus, a price of 10 cents a page (say $20 for a 200-page book) might be obscenely high for a major textbook or trade book, but it might be insanely low for a specialized monograph.

Reference Books. Because science produces prolific data, STM publishers produce a wide variety of compilations of data. Most of these are of the "handbook" variety. Some of the larger fields also have their encyclopedias and dictionaries. Bibliographies were once a common type of reference book, but relatively few are being produced today. (As online bibliographic searching becomes more common, printed bibliographies in most fields will become obsolete.)

Reference books are expensive to produce. Most are produced by commercial publishers, who design the product and employ scientists as consultants to insure the accuracy of the product. The published reference works, particularly the multi-volume works, are likely to be expensive. From the reviewer's point of view, the essential considerations are the usefulness and the accuracy of the data assembled in the work.

Textbooks. STM publishers love textbooks because that is where the money is. A successful undergraduate text in a broad subject may sell tens of thousands of copies. New editions of established texts are published frequently (primarily to kill the competition from the used-book market), and some scientists have become modestly wealthy from textbook royalties.

A textbook is unique in that its success is not determined by its purchasers (students) but by its adopters (professors). Thus, STM publishers try to commission the big names in science to write texts, hoping that major adoptions will result on the basis of name recognition. Occasionally, the big names, who became well known because of their research, write good texts. At least, the science is likely to be first-rate and up to date. Unfortunately, some brilliant and successful researchers are poor writers, and their texts may be almost useless as teaching aids. It shouldn't have to be said but it does: A good reviewer should evaluate a text on the basis of its usefulness as a text; the name on the cover should be irrelevant.

Trade Books. Trade books are those books that are sold primarily through the book trade, that is, book wholesalers and retailers. The typical retail bookstore caters to the tastes of a general audience, those people who walk in off the street. Because a bookstore has space to stock only a small fraction of the total output of publishers, the bookstore is likely to stock only those titles that would interest many potential readers. In bookstores, you will find books that appear on various best-seller lists, popular fiction and general-interest nonfiction, and perhaps not much else.

Bookstores do sell "science" books, however. They sell them by the millions. But these are not the monographs, the reference books, or the textbooks (except in college bookstores). These are the books *about* science written for the general public. Many, unfortunately, are not very scientific, and some are disgustingly pseudoscientific. Have you looked at a best-seller list lately? In the nonfiction category, perhaps half may deal with scientific subjects. Books on nutrition and diet, on health, and on exercise and "fitness" are especially popular in today's market.

Although some of these books are trivial or even a perversion of scientific knowledge, many very good STM books are also sold in bookstores. There are many first-rate books that treat science and scientists in an interesting, educational way. Biographies of prominent scientists seem to find a ready market. Almost all bookstores carry books on everything from the atom to the universe.

Components of the Book Review

Because I believe that there are fundamental differences among the various kinds of STM books, I described them in some detail above.

Now let me go over the same ground to define what should be covered in an effective book review.

Monographs. We can define a monograph as a specialized book written for a specialized audience. Therefore, the reviewer of a monograph has one paramount obligation: to describe for potential readers exactly what is in the book. What, precisely, is the subject of the book, and what are the outside limits of the material covered? If the monograph has a number of subjects, perhaps each with a different author, each subject should be treated individually. The good review, of course, will mirror the quality of the book; the pedestrian material will be passed over quickly, and the significant contributions will be given weightier discussion. The quality of the writing, with rare exceptions, will not need comment. It is the information in the monograph that is important to its audience. Highly technical language, and even some jargon, are to be expected.

Reference Books. The subject of a reference book is likely to be much broader than that of a monograph. Still, it is important for the reviewer to define in appropriate detail the content of the book. Unlike the monograph, which may contain many opinions and other subjective material, the reference book contains facts. Therefore, the prime responsibility of the reviewer is to determine, however possible, the accuracy of the material in the reference book. Any professional librarian will tell you that an inaccurate reference book is worse than none at all.

Textbooks. In reviewing a textbook, the reviewer has a different set of considerations. Unlike the language in a monograph, that in a textbook must be nontechnical and jargon must be avoided. The reader will be a student, not a peer of the scientist who wrote the book. Technical terms will be used, of course, but each should be carefully defined at first use. Unlike the reference book, accuracy is not especially important. An inaccurate number or word here and there is not crucial as long as the *message* gets through. The function of the reviewer, then, is to determine whether the subject of the text is treated clearly, in a way that is likely to enable students to grasp and to appreciate the knowledge presented. The textbook reviewer has one additional responsibility. If other texts on the same subject exist, which is usually the case, the reviewer should provide appropriate comparisons. A new textbook might be "good" based on its own evident merits; however, if it is not as good as existing texts, it is bad.

Trade Books. Again, the reviewer has different responsibilities. The reader of a trade book may be a general reader, not a scientist or a student of the sciences. Therefore, the language must be nontechnical. Furthermore, unlike any of the other STM books, a trade book must be *interesting.* Trade books are bought as much for entertainment as they are for education. Facts may be important, but a boring effusion of facts would be out of place. Scientific precepts are sometimes difficult for the layman to comprehend. The scientist writing for this market must always keep this point in mind, and the reviewer of a trade book must do so also. If a somewhat imprecise, nontechnical term must replace a precise, technical term, so be it. The reviewer may wince from time to time, but a book which succeeds in fairly presenting scientific concepts to the general public should not be faulted because of an occasional imprecision.

Finally, with trade books (as with other STM books, for that matter), the reviewer should try to define the audience. Can any literate person read and understand the book, or is some level of scientific competency necessary?

If a reviewer has done the job well, a potential reader will know whether or not to read the book under consideration, and why.

Chapter 23

How to Write a Thesis

*The average Ph.D. thesis is nothing but
a transference of bones from one graveyard
to another.*

—J. Frank Dobie

Purpose of the Thesis

A Ph.D. thesis in the sciences is supposed to present the candidate's original research. Its purpose is to prove that the candidate is capable of doing original research. Therefore, a proper thesis should be like a scientific paper, which has the same purpose. A thesis should exhibit the same form of disciplined writing that would be required in a journal publication. Unlike the scientific paper, the thesis may describe more than one topic and it may present more than one approach to some topics. The thesis may present all or most of the data obtained in the student's thesis-related research. Therefore, the thesis usually can be longer and more involved than a scientific paper. But the concept that a thesis must be a bulky 200-page tome is wrong, dead wrong. Most 200-page theses I have seen are composed of maybe 50 pages of good science. The other 150 pages comprise turgid descriptions of picayune details.

I have seen a great many Ph.D. theses, and I have assisted with the writing and organization of a good number of them. On the basis of this experience, I have concluded that there are almost no generally accepted rules for thesis preparation. Most types of scientific writing are highly structured. Thesis writing is not. The "right" way to write a thesis varies

124

widely from institution to institution and even from professor to professor within the same department of the same institution.

The dustiest part of most libraries is that area where the departmental theses are shelved. Without doubt, many nuggets of useful knowledge are contained in theses, but who has the time or patience to sort through the hundreds of pages of trivia to find the page or two of useful knowledge?

Reid (32) is one of many who have suggested that the traditional thesis no longer serves a purpose. In Reid's words, "Requirements that a candidate must produce an expansive traditional-style dissertation for a Ph.D. degree in the sciences must be abandoned. ... The expansive traditional dissertation fosters the false impression that a typed record must be preserved of every table, graph, and successful or unsuccessful experimental procedure."

If a thesis serves any real purpose, that purpose might be to determine literacy. Perhaps universities have always worried about what would happen to their image if it turned out that a Ph.D. degree had been awarded to an illiterate. Hence, the thesis requirement. Stated more positively, the candidate has been through a process of maturation, discipline, and scholarship. The "ticket out" is a satisfactory thesis.

It may be useful to mention that theses in European universities are taken much more seriously. They are designed to show that the candidate has reached maturity and can both "do" science and "write" science. Such theses may be submitted after 10 years of work and a number of primary publications, with the thesis itself being a "review paper" that brings it all together.

Tips on Writing

There are few rules for writing a thesis, except those that may exist in your own institution. (The unwritten rule: Write your thesis to please your major professor, if you can figure out what turns him or her on.) If you do not have rules to follow, go to your departmental library and examine the theses submitted by previous graduates of the department, especially those who have gone on to fame and fortune. Perhaps you will be able to detect a common flavor. Whatever ploys worked in the past for others are likely to work for you now.

Generally, a thesis should be written in the style of a review paper. Its purpose is to review the work that led to your degree. Your original data (whether previously published or not) should of course be incorporated, buttressed by all necessary experimental detail. Each of several sections might actually be designed along the lines of a research paper (Introduction, Materials and Methods, Results, Discussion). Overall, however, the parts should fit together like those of a monographic review paper.

Be careful about the headings. If you have one or several Results sections, these must be *your* results, not a mixture of your results with those of others. If you need to present results of others, to show how they confirm or contrast with your own, you should do this within a Discussion section. Otherwise, confusion may result or, worse, you could be charged with lifting data from the published literature.

Start with and work from a carefully prepared outline. In your outline and in your thesis you will of course describe in meticulous detail your own research results. In a thesis, however, it is customary to review all related work. (How else could you pad it out to 150 pages?) Further, there is no bar in a thesis, as there may be in state of the art review papers, to hoary tradition, so it is often desirable to go back into the history of your subject. You might thus compile a really valuable review of the literature of your field, while at the same time learning something about the history of science, which could turn out to be the most valuable part of your education.

I recommend that you give very special attention to the Introduction to your thesis, for two reasons. First, for your own benefit, you need to clarify the problem you attacked, how and why you selected that problem, how you attacked it, and what you learned during the course of your studies. The rest of the thesis should then flow easily and logically from the Introduction. Second, because first impressions are important, you do not want to lose your readers in a cloud of obfuscation right at the outset.

When to Write the Thesis

You would be wise to begin writing your thesis long before it is due. (Try my philosophy: start slowly and then taper off.) In fact, when a particular set of experiments or some major facet of your work has been completed, you should write it up while it is still fresh in your mind. If you save everything until the end, you may find that you have forgotten important details. Worse, you may find that you just don't have time to do a proper writing job. If you have not done much writing previously, you will be amazed at what a painful and time-consuming process it is. You are likely to need a total of 3 months to write the thesis, on a relatively full-time basis. You will not have full time, however, nor can you count on the ready availability of your thesis advisor or your typist. Allow 6 months at a minimum.

Certainly, the publishable portions of your research work should be written as papers and submitted if at all possible before you leave the institution. It will be difficult to do this after you leave the institution, and it will get more difficult with each passing month.

Relationship to the Outside World

Remember, your thesis will bear only your name. Theses are normally copyrighted in the name of the author. Your early reputation and perhaps your job prospects may relate to the quality of your thesis and of the related publications that may appear in the primary literature. A tightly written, coherent thesis will get you off to a good start. An overblown encyclopedia of minutiae will do you no credit. The writers of good theses try hard to avoid the verbose, the tedious, and the trivial.

Be particularly careful in writing the Abstract of your thesis. The Abstracts of theses from most institutions are published in *Dissertation Abstracts*, thus being made available to the larger scientific community.

If your interest in this book at this time centers on how to write a thesis, I suggest that you now carefully read Chapter 20 ("How to Write a Review Paper"), because in many respects a thesis is indeed a review paper.

Chapter 24

How to Present a Paper Orally

I come not, friends, to steal away your hearts:
I am no orator, as Brutus is;
But, as you know me all, a plain, blunt man,
That love my friend.
—WILLIAM SHAKESPEARE

Organization of the Paper

The best way (in my opinion) to organize a paper for oral presentation is to proceed in the same logical pathway that one usually does in writing a paper, starting with "what was the problem?" and ending with "what is the solution?" However, it is important to remember that oral presentation of a paper does *not* constitute publication, and therefore different rules apply. The greatest distinction is that the published paper must contain the full experimental protocol, so that the experiments can be repeated. The oral presentation, however, need not and should not contain all of the experimental detail, unless by chance you have been called upon to administer a soporific at a meeting of insomniacs. Extensive citation of the literature is also undesirable in an oral presentation.

If you will accept my statement that oral presentations should be organized along the same lines as written papers, I need say nothing more about "organization." This material is covered in Chapter 21, "How to Write a Conference Report."

128

The Ten Commandments of Good Speaking

Having had many opportunities to bore audiences myself, and having myself been bored by countless other speakers, I believe that I can now state the fundamental commandments of effective speaking.

1. Be relaxed. Show that you are relaxed by pacing back and forth across the stage.

2. Be informal. Comb your hair or clean your ears from time to time, to maintain a proper air of informality.

3. Be casual. Start by dressing casually. (Loud sport shirts are nice; tank tops are better.) Continue with a casual opening. ("Well, girls and boys, sit back and relax while I lay on some new info fresh from old Mama Nature.")

4. Be memorable. Your audience will be sure to remember you if you successfully develop a memorable characteristic. A pronounced tic would be useful. Make faces. Stare at the ceiling. Close your eyes for extended periods (which will also improve your concentration).

5. Use hand gestures to attract attention. It helps to pretend that you are trying to flag down local aircraft.

6. Use distinctive language. Just as professional athletes discovered the powerful effectiveness of a "you know" added to almost all sentences, scientists too can be remembered for their linguistic abilities. I recall one scientist who became known for his elegant language by the simple device of inserting the word "elegant" into every second sentence.

7. Speak softly. If you rudely awaken the sleepers, they may retaliate by asking nasty questions during the question period.

8. Mumble. Only when your results are absolutely incontrovertible should you speak clearly; otherwise, mumble.

9. Use slides that each illustrate many points. A slide with only one salient point is an insult to the intelligence of the audience.

10. If you have been asked to give a 10-minute presentation, talk for 30 minutes. Only thus will the audience, and especially your colleagues on the panel, be impressed with the vast extent of your knowledge.

Slides

At small, informal scientific meetings, various types of visual aids may be used. Overhead projectors, flip charts, and even blackboards can be used effectively. At most scientific meetings, however, 35-mm slides are the lingua franca. Every scientist *should* know how to prepare effective slides, yet attendance at almost any meeting quickly indicates that many do not.

Here are a few of the considerations that are important. First, slides should be designed specifically for use with oral presentations. Slides prepared from graphs that were drawn for journal publication are seldom effective and often are not even legible. Slides prepared from a type-written manuscript or from a printed journal or book are almost never effective. It should also be remembered that slides should be wide rather than high, which is just the opposite of the preferred dimensions for printed illustrations. Even though 35-mm slides are square (outside measurements of 2 × 2 inches or 50 × 50 mm), the conventional 35-mm camera produces an image area that is 36.3 mm wide and 24.5 mm high. Thus, horizontally oriented slides are usually preferable.

Second, slides should be prepared by professionals or at least by use of professional equipment, such as lettering devices and press-on letters (Letraset, Prestype, etc.). Slides prepared with standard typewriters are almost never effective; the lettering is simply too small. (Use of an IBM Orator typing element, which types in large capital letters, may produce satisfactory results.)

Third, it should be remembered that the lighting in meeting rooms is seldom optimum for slides. Contrast is therefore important. The best slides are made with white lettering on a blue background.

Fourth, slides should not be crowded. Each slide should be designed to illustrate a particular point, or perhaps to summarize a few.

Fifth, get to the hall ahead of the audience. Check the projector, the advance mechanism, and the lights. Make sure that your slides are inserted in the proper order and in proper orientation. There is no need for, and no excuse for, slides that appear out of sequence, upside down, or out of focus.

Normally, each slide should make one simple, easily understood visual statement. The slide should supplement what you are saying at the time the slide is on the screen; the slide should *not* simply repeat what you are saying. And you should *never* read the slide text to the audience.

Slides that are thoughtfully designed and well prepared can greatly enhance the value of a scientific presentation. Poor slides would have ruined Cicero.

The Audience

The presentation of a paper at a scientific meeting is a two-way process. Because the material being communicated at a scientific conference is likely to be the newest available information in that field, both the speakers and the audience should accept certain obligations. As indicated above, speakers should present their material clearly and effectively so that the audience can understand and learn from the information being communicated.

Almost certainly, the audience for an oral presentation will be more diverse than the readership of a scientific paper. Therefore, the oral presentation should be pitched at a more general level than would be a written paper. Avoid technical detail. Define terms. Explain difficult concepts. A bit of redundancy can be very helpful.

For communication to be effective, the audience also has various responsibilities. These start with simple courtesy. The audience should be quiet and attentive. Speakers respond well to an interested, attentive audience, whereas the communication process can be virtually destroyed when the audience is noisy or, worse, asleep.

The best part of an oral presentation is often the question and answer period. During this time, members of the audience have the option, if not the obligation, of raising questions not covered by the speakers, and of briefly presenting ideas or data that confirm or contrast with those presented by the speaker. Such questions and comments should be stated courteously and professionally. This is not the time (although we have all seen it) for some windbag to vent spleen or to describe his or her own erudition in infinite detail. It is all right to disagree, but do not be disagreeable. In short, the speaker has an obligation to be considerate to the audience, and the audience has an obligation to be considerate to the speaker.

Chapter 25

Ethics, Rights, and Permissions

There is a little watchman in my heart
Who is always telling me what time it is.
—EDWIN ARLINGTON ROBINSON

Importance of Originality

In any kind of publishing, various legal and ethical principles must be considered. The principal areas of concern, which are often related, involve originality and ownership (copyright). To avoid charges of plagiarism or copyright infringement, certain types of permission are mandatory if someone else's work, and sometimes even your own, is to be republished.

In science publishing, the ethical side of the question is even more pronounced, because originality in science has a deeper meaning than it does in other fields. A short story, for example, can be reprinted many times without violating ethical principles. A primary research paper, however, can be published in a primary journal only once. Dual publication can be legal if the appropriate copyright release has been obtained, but it is universally considered to be a cardinal sin against the ethics of science. "Repetitive publication of the same data or ideas for different journals, foreign or national, reflects scientific sterility and constitutes exploitation of what is considered an ethical medium for propagandizing one's self. Self-plagiarism signifies lack of scientific objectivity and modesty" (10).

Every primary research journal requires originality, the requirement being usually stated in the journal masthead statement or in the "Instructions to Authors." Typically, such statements read as follows.

"Submission of a paper (other than a review) to a journal normally implies that it presents the results of original research or some new ideas not previously published, that it is not under consideration for publication elsewhere, and that, if accepted, it will not be published elsewhere in the same form, either in English or in any other language, without the consent of the editors" ("General Notes on the Preparation of Scientific Papers," The Royal Society, London).

The "consent of the editors" would not be given if you asked to republish all or a substantial portion of your paper in another primary publication. Even if such consent were somehow obtained, the editor of the second journal would refuse publication if he or she was aware of prior publication. Normally, the consent of the editors (or whoever speaks for the copyright owner) would be granted only if republication were in a nonprimary journal. Obviously, parts of the paper, such as tables and illustrations, could be republished in a review. Even the whole paper could be republished if the nonprimary nature of the publication was apparent; as examples, republication would almost always be permitted in a Collected Reprints volume of a particular institution, in a Selected Papers volume on a particular subject, or in a Festschrift volume comprising papers of a particular scientist. In all such instances, however, appropriate permission should be sought, for both ethical and legal reasons.

Copyright Considerations

The legal reasons relate to copyright law. If a journal is copyrighted, and almost all of them are, legal ownership of the published papers becomes vested in the copyright holder. Thus, if you wish to republish copyrighted material, you must obtain approval of the copyright holder or risk suit for infringement.

Publishers acquire copyright so that they will have the legal basis, acting in their own interests and on behalf of all authors whose work is contained in the journals, for preventing unauthorized use of such published work. Thus, the publishing company and its authors are protected against plagiarism, misappropriation of published data, unauthorized reprinting for advertising and other purposes, and other potential misuse.

Under the 1909 Assignment of Copyright Act, the act of submission of a manuscript to a journal was presumed to carry with it assignment of the author's ownership to the journal (publisher). Upon publication of the journal, with the appropriate copyright imprint in place and followed by the filing of copies and necessary fees with the Register of Copyrights, ownership of all articles contained in the issue effectively passed from the authors to the publisher.

The Copyright Act of 1976, which became effective on 1 January 1978, requires that henceforth this assignment may no longer be assumed; it must be in writing. In the absence of a written transfer of copyright, the publisher is presumed to have acquired only the privilege of publishing the article in the journal itself; the publisher would then lack the right to produce reprints, photocopies, and microfilms or to license others to do so (or to legally prevent others from doing so).

Therefore, most publishers now require that each author contributing to a journal assign copyright to the publisher, either at the time the manuscript is submitted or at the time that it is accepted for publication. To effect this assignment, the publisher provides each submitting author with a document usually titled "Copyright Transfer Agreement."

Another feature of the new Copyright Act that is of interest to authors deals with photocopying. On the one hand, authors wish to see their papers receive wide distribution. On the other hand, they do not (we hope) want this to take place at the expense of the journals. Thus, the new law reflects these conflicting interests by defining as "fair use" certain kinds of library and educational copying (that is, copying that may be done without permission and without payment of royalties), while at the same time protecting the publisher against unauthorized systematic copying.

To make it easy to authorize systematic photocopiers to use journal articles and to remit royalties to publishers, a Copyright Clearance Center has been established. Most scientific publishers of any size have already joined the Center. This central clearinghouse makes it possible for a user to make as many copies as desired, without the necessity of obtaining prior permission, if the user is willing to pay the publisher's stated royalty to the Center. Thus, the user need deal with only one source, rather than facing the necessity of getting permission from and then paying royalties to hundreds of different publishers.

Because both scientific ethics and copyright law are of fundamental importance, every scientist must be acutely sensitive to them. Basically, this means that you must not republish tables, figures, and substantial portions of text *unless* you have acquired permission from the owner of the copyright. Even then, it is important that you label such reprinted materials, usually with a credit line reading "Reprinted with permission from (journal or book reference); copyright (year) by (owner of copyright)."

Of the two (ethics and copyright law), ethics is the more important. Even in flagrant cases of copyright infringement, it might be difficult to prove "damages"; thus, the threat of a lawsuit is more apparent than real. However, when you do not give proper credit to sources, even brief paraphrases of someone else's work can be a violation of the ethics of your profession. Such breaches of ethics, even if unintentional, may adversely affect your standing among your peers.

Simply put, it is the responsibility of every scientist to maintain the integrity of scientific publication.

Incidentally, an ethical manufacturer decided to offer a money-back guarantee. At the bottom of each and every product ad appeared the statement: "If not satisfactory, money will be returned." However, the manufacturer told me that *all* of the money received had been satisfactory.

Chapter 26

Use and Misuse of English

Never fear big words. Long words name little things. All big things have little names, such as Life and death, Peace and war, Or dawn, day, night, love, home. Learn to use little words In a big way—It is hard to do. But they say what you mean. When you don't know what you mean, use big words: They often fool little people.

—SSC BOOKNEWS, July 1981

Keep It Simple

In the earlier chapters of this book, I presented an outline of the various components that could and perhaps should go into a scientific paper. Perhaps, with this outline, the paper won't quite write itself. But if this outline, this table of organization, is followed, I believe that the writing might be a good deal easier than otherwise.

Of course, you still must use the English language. For some of you, this may be difficult. If your native language is not English, you may have a problem. It may not be a big problem, however, if you learn to "keep it simple." You might also benefit from reading Barnes' new book on *Communication Skills for the Foreign-Born Professional* (7). If your native language is English, you still have a problem because the native language of many of your readers is not English.

In other words, keep it simple. Plato had the idea many centuries ago: "Beauty of style and harmony and grace and good rhythm depend on simplicity."

Learn to appreciate, as most managing editors have learned to appreciate, the sheer beauty of the simple declarative sentence. You will then avoid most serious grammatical problems.

Split Infinitives, Dangling Modifiers, and Other Crimes

Most of us these days don't worry about things like split infinitives, but they can be overdone. I quote the best one in my collection, from a legal decision of Judge Thomas, who expunged the Grand Jury Report and upheld the 25 indictments of students and faculty after the Kent State murders. Judge Thomas decided that the Grand Jury Report should be stricken because "it would be unreasonable to expect or ask a prospective juror to honestly to promise to completely disregard these findings and to treat the indictments not as proof of guilt but only as an accusation of crime." On the basis of that sentence alone, I indict Judge Thomas of a far greater crime than anything attributed to the Kent State students.

It is not always easy to recognize a dangling participle or gerund, but you can avoid such faults by giving proper attention to syntax. The word "syntax" refers to that part of grammar dealing with the way in which words are put together to form phrases, clauses, or sentences. According to Will Rogers: "Syntax must be bad, having both *sin* and *tax* in it."

That is not to say that a well-dangled participle or other misplaced modifier isn't a joy to behold, after you have developed a taste for such things. The working day of a managing editor wouldn't be complete until he or she had savored such a morsel as "Lying on top of the intestine, you will perhaps make out a small transparent thread."

Those of you who use chromatographic procedures may be interested in a new technique reported in a manuscript submitted to the *Journal of Bacteriology*: "By filtering thru Whatman no. 1 filter paper, Smith separated the components."

Of course, such charming grammatical errors are not limited to science. I was reading a mystery novel, *Death Has Deep Roots* by Michael Gilbert, when I encountered a particularly sexy misplaced modifier: "He placed at Nap's disposal the marriage bed of his eldest daughter, a knobbed engine of brass and iron."

A Hampshire, England, fire department received a government memorandum seeking statistical information. One of the questions was, "How many people do you employ, broken down by sex?" The fire chief took that question right in stride, answering "None. Our problem here is booze."

If any of you share my interest in harness racing, you may remember that the 1970 Hambletonian was won by a horse named Timothy T. According to *The Washington Post* account of the story, Timothy T. evidently has an interesting background: "Timothy T.—sired by Ayres, the 1964 Hambletonian winner with John Simpson in the sulky—won the first heat going away."

I really like *The Washington Post*. Some time ago they ran an article titled "Antibiotic-Combination Drugs Used to Treat Colds Banned by FDA." Perhaps the next FDA regulation will ban all colds, and virologists will have to find a different line of work.

And, as is well known, the proofreaders at *The Washington Post* have won several Pulitzers. An example of their artistry is the following (from the 1 November 1979 issue of the *Post*):

'Suicide Forest' Toll 43 So Far This Year

Reuter

FUJI-YOSHIDA, Japan, Oct. 31—The bodies of 43 suicides were recovered this year from the infamous "Forest of No Return" at the foot of Mount Fuji near here, police said today.

In the final search of the year, police and firement combed the forest yesterday and found five bodies.

At least 176 bodies have been recovered from the area since 1975.

A novel published in 1960 in Japan June 7. A joke's a joke, but hey, cut called Edwards that same afternoon, covered from the area since 1975.

Joyce Selcnick was not amued. She and later serialized on television glamorized the forest as a place for peaceful death, especially for persons thwarted in love.

Thinking of libraries, I can suggest a new type of acquisition. I once edited a manuscript containing the sentence: "A large mass of literature has accumulated on the cell walls of staphylococci." After the librarians have catalogued the staphylococci, they will have to start on the fish, according to this sentence from a recent manuscript: "The resulting disease has been described in detail in salmon."

A published book review contained this sentence: "This book includes discussion of shock and renal failure in separate chapters."

The first paragraph of a news release issued by the American Lung Association said, " 'Women seem to be smoking more but breathing less,' says Colin R. Woolf, M.D., Professor, Department of Medicine, University of Toronto. He presented evidence that women who smoke are likely to have pulmonary abnormalities and impaired lung function at the annual meeting of the American Lung Association." Even though the ALA meeting was in the lovely city of Montreal, I hope that women who smoke stayed home.

The Ten Commandments of Good Writing

Herewith, I offer "The Ten Commandments of Good Writing":

1. Each pronoun should agree with their antecedent.
2. Just between you and I, case is important.
3. A preposition is a poor word to end a sentence with. (Incidentally, did you hear about the streetwalker who violated a grammatical rule? She unwittingly approached a plainclothesman, and her proposition ended with a sentence.)
4. Verbs has to agree with their subject.
5. Don't use no double negatives.
6. Remember to never split an infinitive.
7. When dangling, don't use participles.
8. Join clauses good, like a conjunction should.
9. Don't write a run-on sentence it is difficult when you got to punctuate it so it makes sense when the reader reads what you wrote.
10. About sentence fragments.

Actually, I have changed my mind about the use of double negatives. During the last Presidential election, I visited my old home town, which is in the middle of a huge cornfield in Northern Illinois. Arriving after a lapse of some years, I was pleased to find that I could still understand the natives. In fact, I was a bit shocked to find that their language was truly expressive even though they were blissfully unaware of the rule against double negatives. One evening at the local gathering place, appropriately named the Farmer's Tavern, I orated at the man on the next stool about the relative demerits of Carter and Reagan. His disinterest was then communicated in the clear statement: "Ain't nobody here knows nothin about politics." While I was savoring this triple negative, a morose gent at the end of the bar looked soulfully into his beer and proclaimed: "Ain't nobody here knows nothin about nothin nohow." Strangely, this quintuple negative said it all.

Metaphorically Speaking

Although this point is not covered by the above rules, I suggest that you watch your similes and metaphors. We have all seen mixed metaphors and noted how comprehension gets mixed along with the metaphor. A rarity along this line is a type that I call the "self-cancelling metaphor." The favorite in my collection was ingeniously concocted by the eminent microbiologist L. Joe Berry. After one of his suggestions had been quickly negated by a committee vote, Joe said, "Boy, I got shot down in flames before I ever got off the ground."

Watch for hackneyed expressions. These are usually similes or metaphors (e.g., timid as a mouse). Interesting and picturesque writing results from the use of fresh similes and metaphors; dull writing results from the use of stale ones.

Some words have become hackneyed, usually by being hopelessly locked to some other word. An example is the verb "wreak." One can "wreak havoc" but nothing else seems to get wreaked these days. Since the dictionary says that "wreak" means "to bring about," one should be able to "wreak a weak pain for a week." To wreak a wry smile, try saying "I've got a weak back." When someone asks when you got it, you respond "Oh, about a week back." (At the local deli, we call this tongue in cheek on wry.)

Misuse of Words

Also watch for self-cancelling or redundant words. I recently heard someone described as being a "well-seasoned novice." A newspaper article referred to "young juveniles." A sign in a stamp and coin dealer's shop read "authentic replicas." If there is any expression that is dumber than "7 a.m. in the morning," it is "viable alternative." (If an alternative is not viable, it is not an alternative.)

Certain words are wrongly used thousands of times in scientific writing. If someone were to draw up a list of the worst offenders, I would nominate the following:

amount. Use this word when you refer to a mass or aggregate. Use number when units are involved. "An amount of cash" is all right. "An amount of coins" is wrong.

and/or. This is a slipshod construction used by thousands of authors but accepted by few experienced editors. Bernstein (8) said, "Whatever its uses in legal or commercial English, this combination is a visual and mental monstrosity that should be avoided in other kinds of writing."

case. This is the most common word in the language of jargon. Better and shorter usage should be substituted: "in this case" means "here"; "in most cases" means "usually"; "in all cases" means "always"; "in no case" means "never."

each-every. If I had a dollar for every mistake I have made, how much would I have? The answer is one dollar. If I had a dollar for each mistake I have made, I would be a millionaire.

it. This common, useful pronoun can cause a problem if the antecedent is not clear, as in the sign which read: "Free information about VD. To get it, call 654—7000."

like. Often used incorrectly as a conjunction. Should be used only as a preposition. When a conjunction is needed, substitute "as." Like I just said, this sentence should have started with *As.*

only. Many sentences are only partially comprehensible because the word *only* is positioned correctly in the sentence only some of the time. Consider this sentence: "I hit him in the eye yesterday." The word *only* can go between any two words in the sentence, but look at the differences in meaning that result.

quite. This word is often used in scientific writing. Next time you notice it in one of your manuscripts, delete the word and read the sentence again. You will notice that, without exception, *quite* is quite unnecessary.

varying. The word means "changing." Often used erroneously when "various" is meant. "Various concentrations" are defined concentrations that do not vary.

which. Although "which" and "that" can often be used interchangeably, sometimes they cannot. The word "which" is properly used in a "nonrestrictive" sense, to introduce a clause that is not essential to the rest of the sentence; "that" introduces an essential clause. Examine these two sentences: "CetB mutants, *which* are tolerant to colicin E2, also have an altered . . . " "CetB mutants *that* are tolerant to colicin E2 also have an altered . . . " Note the substantial difference in meaning. The first sentence indicates that *all* CetB mutants are tolerant to colicin; the second sentence indicates that only some of the CetB mutants are tolerant to colicin.

while. When a temporal relationship exists, "while" is correct; otherwise, "whereas" would be a better choice. "Nero fiddled while Rome burned" is fine. "Nero fiddled while I wrote a book on scientific writing" is not.

Misuse of words can sometimes be entertaining, if not enlightening. I have always enjoyed the word "thunderstruck," although I have never had the pleasure of meeting anyone who has been struck by thunder. Jimmy Durante built his comedy style around malapropisms. We all enjoy them, but seldom do they contribute to comprehension. Rarely, you might use a malapropism by design, to add picturesque interest to your speaking or writing. One that I have used several times is the classic "I'm really nostalgic about the future."

This reminds me of the story about a graduate student recently arrived in this country from one of the more remote countries of the world. He had a massive English vocabulary, developed by many years of assiduous study. Unfortunately, he had had few opportunities to speak the language. Soon after his arrival in this country, the dean of the school invited a number of the students and faculty to an afternoon tea. Some of the faculty members soon engaged the new foreign student in conversation. One of the first questions asked was "Are you married?" The student said, "Oh, yes, I am most entrancingly married to one of the most exquisite belles of my country, who will soon be arriving here in the United States, ending our temporary bifurcation." The faculty members exchanged questioning glances—then came the next question: "Do you have children?" The student answered "No." After some thought, the student decided this answer needed some amplification, so he said, "You see, my wife is inconceivable." At this, his questioners could not hide their smiles, so the student, realizing he had committed a faux pas, decided to try again. He said, "Perhaps I should have said that my wife is impregnable." When this comment was greeted with open laughter, the student decided to try one more time: "I guess I should have said my wife is unbearable."

All seriousness aside, is there something about the use (rather than abuse) of English in scientific writing that merits special comment?

Tense in Scientific Writing

There is one special convention of writing scientific papers that is very tricky. It has to do with *tense*, and it is important because proper usage derives from scientific ethics.

When a scientific paper has been validly published in a primary journal, it thereby becomes knowledge. Therefore, whenever you quote previously published work, ethics requires you to treat that work with respect. You do this by using the *present* tense. It is correct to say "Streptomycin inhibits the growth of *M. tuberculosis* (13)." Whenever you quote or discuss previously published work, you should use the present tense; you are quoting established knowledge.

Your own present work must be referred to in the *past* tense. Your work is not presumed to be established knowledge until *after* it has been published. If you determined that the optimal growth temperature for *Streptomyces everycolor* was 37°C, you should say "*S. everycolor* grew best at 37°C." If you are citing previous work, possibly your own, it is then correct to say "*S. everycolor* grows best at 37°C."

In the typical paper, you will normally go back and forth between the past and present tenses. Most of the Abstract should be in the past tense, because you are referring to your own present results. Likewise, the Materials and Methods and the Results sections should be in the past tense, as you describe what you did and what you found. On the other hand, most of the Introduction and much of the Discussion should be in the present tense, because these sections usually emphasize previously established knowledge.

Suppose that your research concerned the effect of streptomycin on *Streptomyces everycolor*. The tense would vary somewhat as follows.

In the Abstract, you would say "The effect of streptomycin on *S. everycolor* grown in various media *was* tested. Growth of *S. everycolor*, measured in terms of optical density, *was* inhibited in all media tested. Inhibition *was* most pronounced at high pH levels."

In the Introduction, typical sentences might be "Streptomycin *is* an antibiotic produced by *Streptomyces griseus* (13). This antibiotic *inhibits* the growth of certain other strains of *Streptomyces* (7, 14, 17). The effect of streptomycin on *S. everycolor is* reported in this paper."

In the Materials and Methods section, you would say "The effect of streptomycin *was* tested against *S. everycolor* grown on Trypticase soy agar (BBL) and several other media (Table 1). Various growth temperatures and pH levels *were* employed. Growth *was* measured in terms of optical density (Klett units)."

In the Results, you would say "Growth of *S. everycolor was* inhibited by streptomycin at all concentrations tested (Table 2) and at all pH levels (Table 3). Maximum inhibition *occurred* at pH 8.2; inhibition *was* slight below pH 7."

In the Discussion, you might say "*S. everycolor was* most susceptible to streptomycin at pH 8.2, whereas *S. nocolor is* most susceptible at pH 7.6 (13). Various other *Streptomyces* species *are* most susceptible to streptomycin at even lower pH levels (6, 9, 17)."

In short, you should normally use the present tense when you refer to previously published work, and you should use the past tense when referring to your present results.

The principal exception to this rule is in the area of attribution and presentation. It is correct to say "Smith (9) *showed* that streptomycin inhibits *S. nocolor*." It is also correct to say "Table 4 *shows* that streptomycin inhibited *S. everycolor* at all pH levels." Another exception is that the results of calculations and statistical analyses should be in the present tense, even though statements about the objects to which they refer are in the past tense; e.g., "These values *are* significantly greater than those of the females of the same age, indicating that the males *grew* more rapidly."

Active Versus Passive Voice

Let us now talk about *voice*. In any type of writing, the active voice is usually more precise and less wordy than the passive voice. (This is not always true; if it were, we would have an Eleventh Commandment: "The passive voice should never be used.") Why, then, do scientists insist on using the passive voice? Perhaps this bad habit is the result of the erroneous idea that it is somehow impolite to use first-person pronouns. As a result, the scientist typically uses such verbose (and imprecise) statements as "It was found that" in preference to the short, unambiguous "I found."

I herewith ask all young scientists to renounce the false modesty of previous generations of scientists. Do not be afraid to name the agent of the action in a sentence, even when it is "I" or "we." Once you get into the habit of saying "I found," you will also find that you have a tendency to write "*S. aureus* produced lactate" rather than "Lactate was produced by *S. aureus*." (Note that the "active" statement is in three words; the passive requires five.)

You can avoid the passive voice by saying "The authors found" instead of "it was found." Compared with the simple "we," however, "the authors" is pretentious, verbose, and imprecise (which authors?).

Singulars and Plurals

If you use first-person pronouns, use both the singular and the plural forms as needed. Do not use the "editorial we" in place of "I." The use of "we" by a single author is outrageously pedantic.

One of the most frequent errors committed in scientific papers is the use of plural forms of verbs when the singular forms would be correct. For example, you should say "10 g *was* added," not "10 g *were* added." This is because a *single* quantity was added. Only if the 10 g were added 1 g at a time would it be correct to say "10 g were added."

The singular-plural problem also applies to nouns. The problem is severe in scientific writing, especially in biology, because so many of our words are, or are derived from, Latin. Most of these words retain their Latin plurals; at least they do when used by careful writers.

Many of these words (e.g., data, media) have entered popular speech, where the Latin "a" plural ending is simply not recognized as a plural. Most people habitually use "data is" constructions and probably have never used the real singular, *datum*. Unfortunately, this lax usage has become so common outside science that even some dictionaries tolerate it. *Webster's New Collegiate Dictionary*, for example, gives "the data is plentiful" as an example of accepted usage. The "careful writer" (8), however, says that "The use of *data* as if it were a singular noun is a common solecism."

This "plural" problem was commented upon by Sir Ashley Miles, the eminent microbiologist and scholar of The London Hospital Medical

College in a letter to me as Editor of *ASM News* (*44*:600, 1978):

> *A Memoranda on Bacterial Motility.* The motility of a bacteria is a phenomena receiving much attention, especially in relation to the structure of a flagella and the effect on it of an antisera. No single explanatory data is available; no one criteria of proof is recognized; even the best media to use is unknown; and no survey of the various levels of scientific approach indicates any one strata, or the several stratae, from which answers may emerge. Flagellae are just as puzzling as the bacteriae which carry them.

Noun Problems

Another frequent problem in scientific writing is the verbosity that results from use of abstract nouns. This malady is corrected by turning the nouns into verbs. "Examination of the patients was carried out" should be changed to the more direct "I examined the patients"; "separation of the compounds was accomplished" can be changed to "the compounds were separated"; "transformation of the equations was achieved" can be changed to "the equations were transformed."

Another problem with nouns results from using them as adjectives. Normally, there is no problem with such usage, but you should watch for special problems. We have no problem with "liver disease" (even though the adjective "hepatic" could be substituted for the noun "liver"). The problem aspect is illustrated by the following sentences from my biography: "When I was 10 years old, my parents sent me to a child psychiatrist. I went for a year and a half. The kid didn't help me at all." I once saw an ad (in *The New York Times*, of all places) with the headline "Good News for Home Sewers." I don't recall whether it was an ad for a drain-cleaning compound or needle and thread. (Did you ever sew a home?)

The problem gets still worse when clusters of nouns are used as adjectives, especially when a real adjective gets into the brew. "Tissue culture response" is awkward; "infected tissue culture response" is incomprehensible (unless responses can be infected).

Appendix 3 lists certain words and expressions, commonly seen in scientific writing, that are often misspelled or misused. You will impress journal editors, and perhaps your family and friends, if you stop committing any obvious spelling and grammatical errors that may have previously characterized your speech and writing.

Odds and Ends

Apropos of nothing, I would mention that English is a strange language. Isn't it curious that the past tense of "have" ("had") is converted to the past participle simply by repetition: He *had had* a serious illness. Strangely, it is possible to string together 11 "hads" in a row in a grammatically correct sentence. If one were to describe a teacher's reaction to themes turned in by students John and Jim, one could say: John, where Jim had had "had," had had "had had"; "had had" had had an unusual effect on the teacher. That peculiar word "that" can also be strung together, as in this sentence: He said, in speaking of the word "that," that that "that" that that student referred to was not that "that" that that other student referred to.

The "hads" and the "thats" in a row show the power of punctuation. As a further illustration, I now mention a little grammatical parlor game that you might want to try on your friends. Hand a slip of paper to each person in the group and ask the members of the group to provide any punctuation necessary to the following seven-word sentence: "Woman without her man is a savage." The average male chauvinist will quickly respond that the sentence needs no punctuation, and he is correct. There will be a few pedants among the male chauvinists who will place balancing commas around the prepositional phrase: "Woman, without her man, is a savage." Grammatically, this is also correct. The truly liberated woman, however, and an occasional liberated man, will place a dash after "woman" and a comma after "her." Then we have "Woman—without her, man is a savage."

Let me end where I started by again emphasizing the importance of syntax. Whenever comprehension goes out the window, faulty syntax is usually responsible. Sometimes, faulty syntax is simply funny and comprehension is not lost, as in these two items, culled from want ads: "For sale, fine German Shepherd dog, obedient, well trained, will eat anything, very fond of children." "For sale, fine grand piano, by a lady, with three legs."

But look at this sentence, which is similar to thousands that have appeared in the scientific literature: "Thymic humoral factor (THF) is a single heat-stable polypeptide isolated from calf thymus composed of 31 amino acids with molecular weight of 3,200." The double prepositional phrase "with molecular weight of 3,200" would logically modify the preceding noun "acids," meaning that the amino acids had a molecular weight of 3,200. Less logically, perhaps the calf thymus had a molecular weight of 3,200. Least logical of all (because of their distance apart in the sentence) would be for the THF to have a molecular weight of 3,200, but, indeed, that was what the author was trying to tell us.

If you have any interest whatsoever in learning to use English effectively, you should read Strunk and White's *The Elements of Style*. The "elements" are given briefly (in 78 pages!) and clearly. Anyone writing anything should read and use this famous "little book."

Chapter 27

Avoiding Jargon

It is clever, of course, to be clever
And good, of course, to be good
But when you're so frightfully clever
As seldom to be understood
It's sad, though if anything sadder
Not to be quite as good as you should.
—LEWIS CARROLL

Mumblespeak and Other Sins

The favorite type of verbosity that afflicts authors is jargon. This syndrome is characterized, in extreme cases, by the total omission of one-syllable words. Writers with this affliction never *use* anything—they *utilize*. They never *do*—they *perform*. They never *start*—they *initiate*. They never *end*—they *finalize* (or *terminate*). They never *make*—they *fabricate*. They use *initial* for *first*, *ultimate* for *last*, *prior to* for *before*, *subsequent to* for *after*, *militate against* for *prohibit*, *sufficient* for *enough*, and *plethora* for *too much*. An occasional author will slip and use the word *drug*, but most will salivate like Pavlov's dogs in anticipation of using *chemotherapeutic agent*. Who would use the three-letter word *now* when they can use the elegant expression *at this point in time*?

Stuart Chase (12) tells the story of the plumber who wrote to the Bureau of Standards saying he had found hydrochloric acid good for cleaning out clogged drains. The Bureau wrote back "The efficacy of hydrochloric acid is indisputable, but the chlorine residue is incompatible with metallic permanence." The plumber replied that he was glad the Bureau agreed. The Bureau tried again, writing "We cannot assume responsibility for the production of toxic and noxious residues with hydrochloric acid, and suggest that you use an alternate procedure." The plumber again said that he was glad the Bureau agreed with him. Finally, the Bureau wrote to the plumber "Don't use hydrochloric acid; it eats hell out of the pipes."

Should we liken the scientist to a plumber, or is the scientist perhaps more exalted? With that Doctor of Philosophy degree, should the scientist know some philosophy? I agree with John W. Gardner, who said, "The society which scorns excellence in plumbing because plumbing is a humble activity and tolerates shoddiness in philosophy because it is an exalted activity will have neither good plumbing nor good philosophy. Neither its pipes nor its theories will hold water" (Science News, p. 137, 2 March 1974).

I like the way that Aaronson (1) put it: "But too often the jargon of scientific specialists is like political rhetoric and bureaucratic mumblespeak: ugly-sounding, difficult to understand, and clumsy. Those who use it often do so because they prefer pretentious, abstract words to simple, concrete ones."

The trouble with jargon is that it is a special language, the meaning of which is known only to a specialized "in" group. Science should be universal, and therefore every scientific paper should be written in a universal language; the use of Sanskrit, Tagalog, and jargon should be strongly discouraged.

Perhaps Theodore Roosevelt had a more jingoistic purpose in mind when he composed the following sentence in a letter read at the All-American Festival, New York, 5 January 1919, but his thought exactly fits scientific writing: "We have room for but one language here, and that is the English language, for we intend to see that the crucible turns our people out as Americans, and not as dwellers in a polyglot boarding house."

Because I believe strongly that the temple of science should not be a polyglot boarding house, I believe that every scientist should avoid jargon. Avoid it not sometimes; avoid it all the time.

Of course, you will have to use specialized terminology on occasion. If such terminology is readily understandable to practitioners and students in the field, there is no problem. If the terminology is *not* recognizable to any portion of your potential audience, you should (i) use more standardized terminology or (ii) carefully define the esoteric terms (jargon) that you are using. In short, you should not write for the half-dozen or so people who are doing exactly your kind of work. You should write for the hundreds of people whose work is only slightly related to yours but who may want or need to know some particular aspect of your work.

Definition of Jargon

According to dictionary definition (e.g., *Webster's New Collegiate Dictionary*), there are three definitions of jargon: "(1) confused, unintelligible language; strange, outlandish, or barbarous language; (2) technical terminology or characteristic idiom of a special activity or group; (3) obscure and often pretentious language marked by circumlocutions and long words."

All three types of jargon should be avoided if possible. The usage described in the first and third definitions should always be avoided. The second definition ("technical terminology") is much more difficult to avoid in scientific writing, but accomplished writers have learned that technical terminology can be used *after* it has been defined or explained. Obviously, you are writing for a technically trained audience; it is only the unusual technical terms that need explanation.

Here are a few important concepts that all readers of this book should master. They are, however, expressed in typical scientific jargon. With a little effort you can probably translate these sentences into simple English:

1. As a case in point, other authorities have proposed that slumbering canines are best left in a recumbent position.

2. It has been posited that a high degree of curiosity proved lethal to a feline.

3. There is a large body of experimental evidence which clearly indicates that smaller members of the genus *Mus* tend to engage in recreational activity while the feline is remote from the locale.

4. From time immemorial, it has been known that the ingestion of an "apple" (i.e., the pome fruit of any tree of the genus *Malus*, said fruit being usually round in shape and red, yellow, or greenish in color) on a diurnal basis will with absolute certainty keep a primary member of the health care establishment absent from one's local environment.

5. Even with the most sophisticated experimental protocol, it is exceedingly unlikely that you can instill in a superannuated canine the capacity to perform novel feats of legerdemain.

6. A sedimentary conglomerate in motion down a declivity gains no addition of mossy material.

7. The resultant experimental data indicate that there is no utility in belaboring a deceased equine.

Bureaucratese

Regrettably, too much scientific writing fits the first and third definitions of jargon. Perhaps because science and scientists are both imitative and institutionalized, the writing of scientists is often like that of government bureaucrats. All too often, scientists write like the legendary Henry B. Quill, the bureaucrat described by Meyer (24): "Quill had mastered the mother tongue of government. He smothered his verbs, camouflaged his subjects and hid everything in an undergrowth of modifiers. He braided, beaded and fringed, giving elaborate expression to negligible thoughts, weasling [*sic*], hedging and announcing the obvious. He spread generality like flood waters in a long, low valley. He sprinkled everything with aspects, feasibilities, alternatives, effectuations, analyzations, maximizations, implementations, contraindications and appurtenances. At his best, complete immobility set in, lasting sometimes for dozens of pages."

Some jargon, or bureaucratese, is made up of clear, simple words, but, when the words are strung together in seemingly endless profusion, their meaning is not readily evident. Examine the following, an important federal regulation (Code of Federal Regulations, Title 36, Paragraph 50.10) designed to protect trees from injury; this notice was posted in

Reprinted by premission of Tribune Company Syndicate, Inc.

National Capital Parks and Planning Commission recreation areas in the Washington area:

TREES, SHRUBS, PLANTS, GRASS AND OTHER VEGETATION

(a) General Injury. No person shall prune, cut, carry away, pull up, dig, fell, bore, chop, saw, chip, pick, move, sever, climb, molest, take, break, deface, destroy, set fire to, burn, scorch, carve, paint, mark, or in any manner interfere with, tamper, mutilate, misuse, disturb or damage any tree, shrub, plant, grass, flower, or part thereof, nor shall any person permit any chemical, whether solid, fluid or gaseous to seep, drip, drain or be emptied, sprayed, dusted or injected upon, about or into any tree, shrub, plant, grass, flower or part thereof except when specifically authorized by competent authority; nor shall any person build fires or station or use any tar kettle, heater, road roller or other engine within an area covered by this part in such a manner that the vapor, fumes or heat therefrom may injure any tree or other vegetation.

(TRANSLATION: Don't mess with growing things.)

Jargon does not necessarily involve the use of specialized words. Faced with a choice of two words, the jargonist always selects the longer one. The jargonist really gets his jollies, however, by turning short, simple statements into a long string of words. And, usually, the longer word or the longer series of words is not as clear as the simpler expression. I challenge anyone to show how "at this point in time" means, in its cumbersome way, more than the simple word "now." The concept denoted by "if" is not improved by substituting the pompous expression "in the event that."

Special Cases

Perhaps the worst offender of all is the word "case." There is no problem with a case of canned goods or even a case of flu. However, 99% of the uses of "case" are jargon. In case you think that 99% is too high, make your own study. Even if my percentage is too high, a good case could be made for the fact that "case" is used in too many cases.

Another word that I find offensive (in all cases) is the word "interface." As far as I know, the only time people can interface is when they kiss.

Still another word that causes trouble (in some cases) is "about," not because it is used but because it is avoided. As pointed out by Weiss (39), writers seem unwilling to use the clear, plain "about" and instead use wordier and less-clear substitutes such as:

approximately	pursuant to
in connection with	re
in reference to	reference
in relation to	regarding
in the matter of	relative to
in the range of	relating to the subject matter of
in the vicinity of	respecting
more or less	with regard to
on the order of	with respect to
on the subject of	within the ballpark of

In Appendix 4 I have collected a few "Words and Expressions to Avoid." A similar list well worth consulting was published by O'Connor and Woodford (28). It is not necessarily improper to use any of these words or expressions *on occasion*; if you use them repeatedly, however, you are writing in jargon and your reader is suffering. "Clutter is the disease of American writing. We are a society strangling in unnecessary words, circular constructions, pompous frills and meaningless jargon" (41).

Perhaps the most common way of creating a new word is the jargonist's habit of turning nouns into verbs. A classic example appeared in a manuscript which read: "One risks exposure when swimming in ponds or streams near which cattle have been pasturized." The copy editor, knowing that there is no such word as "pasturized," changed it to "pasteurized." (I see nothing wrong with that. If you can pasteurize milk, I presume that you can pasteurize the original container.)

In their own pastures, scientists are, of course, very expert, but they often succumb to pedantic, jargonistic, and useless expressions, telling the reader more than the reader wants or needs to know. As the English novelist, George Eliot, said: "Blessed is the man who, having nothing to say, abstains from giving us wordy evidence of this fact."

I'm reminded of the two adventuresome hot-air balloonists who, slowly descending after a long trip on a cloudy day, looked at the terrain below and had not the faintest idea where they were. It so happens that they were drifting over the grounds of one of our more famous scientific research institutes. When the balloonists saw a man walking along the side of a road, one called out, "Hey, mister, where are we?" The man looked up, took in the situation, and, after a few moments of reflection, said, "You're in a hot-air balloon." One balloonist turned to the other and said, "I'll bet that man is a scientist." The other balloonist said, "What makes you think so?" To which the first replied, "His answer is perfectly accurate—and totally useless."

Chapter 28

How and When to Use Abbreviations

Ex Umbris et Imaginibus in Veritatem!
(From shadows and symbols into the truth.)
—JOHN HENRY, CARDINAL NEWMAN

General Principles

Many experienced editors loathe abbreviations. Some editors would prefer that they not be used at all, except for standard units of measurement and their SI (Système International) prefixes, abbreviations for which are allowed in all journals. In your own writing, you would be wise to keep abbreviations to a minimum. The editor will look more kindly on your paper, and the readers of your paper will bless you forever. I doubt that more preaching on this point is necessary because, by now, you yourself have no doubt come across undefined and indecipherable abbreviations in the literature. Just remember how annoyed you felt when you were faced with these conundrums, and join with me now in a vow never again to pollute the scientific literature with an undefined abbreviation.

The "how to" of using abbreviations is easy, because most journals use the same convention. When you plan to use an abbreviation, you introduce it by spelling out the word or term first, followed by the abbreviation within parentheses. The first sentence of the Introduction of a paper might read: "Bacterial plasmids, as autonomously replicating deoxyribonucleic acid (DNA) molecules of modest size, are promising models for studying DNA replication and its control."

The "when to" of using abbreviations is much more difficult. Several general guidelines might be helpful.

First, never use an abbreviation in the title of an article. Very few journals allow abbreviations in titles, and their use is strongly discouraged by the indexing and abstracting services. If the abbreviation is not a standard one, the literature retrieval services will have a difficult or impossible problem. Even if the abbreviation is standard, indexing and other problems arise. One major problem is that accepted abbreviations have a habit of changing; today's abbreviations may be unrecognizable a few years from today. Comparison of certain abbreviations as listed in the various editions of the *Council of Biology Editors Style Manual* emphasizes this point. Dramatic changes occur when the terminology itself changes. Students today could have trouble with the abbreviation "DPN" (which stands for "diphosphopyridine nucleotide"), because the name itself has changed to "nicotinamide adenine dinucleotide," the abbreviation for which is "NAD."

Abbreviations should almost never be used in the Abstract. Only if you use the same name, a long one, quite a number of times should you consider an abbreviation. If you use an abbreviation, you must define it at the first use in the Abstract. Remember that the Abstract will stand alone in whichever abstracting publications cover the journal in which your paper appears.

In the text itself, abbreviations may be used. They serve a purpose in reducing printing costs, by somewhat shortening the paper. More importantly, they aid the reader when they are used judiciously.

Good Practice

It is good practice, when writing the first draft of the manuscript, to spell out all terms. Then examine the manuscript for repetition of long words or phrases that might be candidates for abbreviation. Do not abbreviate a term that is used only a few times in the paper. If the term is used with modest frequency—let us say between three and six times—and a standard abbreviation for that term exists, introduce and use the abbreviation. (Some journals allow some standard abbreviations to be used without definition at first use.) If no standard abbreviation exists, do not manufacture one unless the term is used frequently or is a very long and cumbersome term that really cries out for abbreviation.

Often you can avoid abbreviations by using the appropriate pronoun (it, they, them) if the antecedent is clear, or by using a substitute expression such as "the inhibitor," "the substrate," "the drug," "the enzyme," or "the acid."

Usually, you should introduce your abbreviations one by one as they first occur in the text. Alternatively, you might consider a separate paragraph (headed "Abbreviations Used") in the Introduction or in Materials and Methods. The latter system is especially useful if related reagents, such as a group of organic chemicals, are to be used in abbreviated form later in the paper.

Units of Measurement

Units of measurement are abbreviated when used with numerical values. You would say "4 mg was added." (The same abbreviation is used for the singular and the plural.) When used without numerals, however, units of measurement are not abbreviated. You would say "Specific activity is expressed as micrograms of adenosine triphosphate incorporated per milligram of protein per hour."

Special Problems

A frequent problem with abbreviations concerns use of "a" or "an." Should you say "a M.S. degree" or "an M.S. degree"? Recall the old rule that you use "a" with words beginning with a consonant sound and "an" with words beginning with a vowel sound (e.g., the letter "em"). Because in science we should use only common abbreviations, those not needing to be spelled out in the reader's mind, the proper choice of article should relate to the sound of the first letter of the abbreviation, not the sound of the first letter of the spelled out term. Thus, although it is correct to say, "a Master of Science degree," it is incorrect to say "a M.S. degree." Because the reader reads "M.S." as "em ess," the proper construction is "an M.S. degree."

In biology, it is customary to abbreviate generic names of organisms after first use. At first use, you would spell our "*Streptomyces griseus.*" In later usage, you can abbreviate the genus name but not the specific epithet: *S. griseus.* Suppose, however, that you are writing a paper that concerns species of both *Streptomyces* and *Staphylococcus.* You would then spell out the genus names repeatedly. Otherwise, readers might be confused as to whether a particular "*S.*" abbreviation referred to one genus or the other.

SI Units

Appendix 5 gives the abbreviations for the prefixes used with all SI (Système International) units. The SI units and symbols, and certain derived SI units, have become part of the language of science. This modern metric system should be mastered by all students of the sciences.

The *CBE Style Manual* (11) is a good source for more complete information.

Briefly, SI units include three classes of units: base units, supplementary units, and derived units. The seven base units that form the foundation of SI are the metre, kilogram, second, ampere, kelvin, mole, and candela. In addition to these seven base units, there are two supplementary units for plane and solid angles: the radian and steradian, respectively. Derived units are expressed algebraically in terms of base units or supplementary units. For some of the derived SI units, special names and symbols exist. (The SI units are "metre" and "litre"; the U.S. Bureau of Standards, followed by the American Chemical Society and a number of other publishers, is tenaciously retaining the traditional American spellings, "meter" and "liter.")

Other Abbreviations

Appendix 6 provides a list of acceptable abbreviations that are now considered to be standard. Most of them are straight out of the *CBE Style Manual* (11) or some other respected source. Use these abbreviations when necessary. Avoid most others. Those that you use should be introduced as carefully as you would introduce royalty.

Chapter 29

A Personalized Summary

Perhaps it may turn out a sang,
Perhaps turn out a sermon.
                    —ROBERT BURNS

I have been associated with scientific books and journals for more than a quarter of a century. This experience may have instilled in me a tad or two of wisdom somewhere along the line; certainly, it has instilled prejudices, some of them strong ones. What has been instilled in me will now be distilled and dispensed to you. I leave it to you, the reader, to determine whether this philosophical musing is "sang," sermon, or summary, or none of the above.

Through the years, I have had many occasions to visit various scientific laboratories. Almost always, I have been impressed, sometimes awed, by the obvious quality of the laboratories themselves and of the equipment and supplies they contain. Judging by appearances, one could only believe that the newest and best (and most expensive) instruments and reagents were used in these laboratories.

During those same years, I have seen thousands of the products of those same laboratories. Some of these products (scientific papers) properly reflected the quality and expense that went into their generation. But many did not.

I want to talk about the many that did not. I ask you, as I have often asked myself, why it is that so many scientists, while capable of brilliant performance in the laboratory, write papers that would be given failing marks in a seventh-grade composition class. I ask you why it is that

157

some scientists will demand the newest ultracentrifuge, even if it costs $40,000, and then refuse to pay $20 to a commercial artist to draw a proper graph of the results obtained with the ultracentrifuge. About a dozen similar questions leap to my mind. Unfortunately, I do not know the answers and I doubt that anyone does.

Perhaps there are no "answers." If there are no answers, that leaves me free to do a little philosophizing. (If you have gotten this far in this book, you can heroically hang on for another few paragraphs.)

If we view knowledge as the house we live in, scientific knowledge will tell us how to construct our house. But we need artistic knowledge to make our house beautiful, and we need humanistic knowledge so that we can understand and appreciate life within our house.

If we view a scientific paper as the culmination of scientific research, which it is, we *can*, if we but try, make it more beautiful and more understandable; we can do this by enriching our scientific knowledge with a bit of the arts and humanities. A well-written scientific paper is the product of a well-trained scientist, yes; but the scientist capable of writing a really good paper will also be a cultured man or woman.

Students of the sciences must not content themselves with study of the sciences alone; science will be more meaningful if studied against a background of other knowledge.

Especially, students must learn how to write, because science demands written expression. Erudition is valued in science; unfortunately, it is often equated with long words, rare words, and complex statements. To learn to write, you must learn to read. To learn to write well, you should read good writing. Read your professional journals, but also read Shakespeare.

Many universities now provide courses in scientific writing. Those that do not should be ashamed of themselves.

What I have said in this book is this: Scientific research is not complete until the results have been published. Therefore, a scientific paper is an *essential* part of the research process. Therefore, the writing of an accurate, understandable paper is just as important as the research itself. Therefore, the words in the paper should be weighed as carefully as the reagents in the laboratory. Therefore, the scientist must know how to use words. Therefore, the education of a scientist is not complete until the ability to publish has been established.

APPENDIX 1

List of Journal Title Word Abbreviations

Word	Abbreviation	Word	Abbreviation
Abstracts	Abstr.	Chemical	Chem.
Academy	Acad.	Chemie	Chem.
Acta	No abbrev.	Chemistry	Chem.
Advances	Adv.	Chemotherapy	Chemother.
Agricultural	Agric.	Chimie	Chim.
American	Am.	Clinical	Clin.
Anales	An.	Commonwealth	Commw.
Analytical	Anal.	Comptes	C.
Anatomical	Anat.	Conference	Conf.
Annalen	Ann.	Contributions	Contrib.
Annales	Ann.	Current	Curr.
Annals	Ann.	Dairy	No abbrev.
Annual	Annu.	Dental	Dent.
Anthropological	Anthropol.	Developmental	Dev.
Antibiotic	Antibiot.	Diseases	Dis.
Antimicrobial	Antimicrob.	Drug	No abbrev.
Applied	Appl.	Ecology	Ecol.
Arbeiten	Arb.	Economics	Econ.
Archiv	Arch.	Edition	Ed.
Archives	Arch.	Electric	Electr.
Archivio	Arch.	Electrical	Electr.
Association	Assoc.	Engineering	Eng.
Astronomical	Astron.	Entomologia	Entomol.
Atomic	At.	Entomologica	Entomol.
Bacteriological	Bacteriol.	Entomological	Entomol.
Bacteriology	Bacteriol.	Environmental	Environ.
Bakteriologie	Bakteriol.	Ergebnisse	Ergeb.
Berichte	Ber.	Ethnology	Ethnol.
Biochemical	Biochem.	European	Eur.
Biochimica	Biochim.	Excerpta	No abbrev.
Biological	Biol.	Experimental	Exp.
Biologie	Biol.	Fauna	No abbrev.
Botanical	Bot.	Federal	Fed.
Botanisches	Bot.	Federation	Fed.
Botany	Bot.	Fish	No abbrev.
British	Br.	Fisheries	Fish.
Bulletin	Bull.	Flora	No abbrev.
Bureau	Bur.	Folia	No abbrev.
Cell	No abbrev.	Food	No abbrev.
Cellular	Cell.	Forest	For.
Central	Cent.	Forschung	Forsch.

Word	Abbreviation	Word	Abbreviation
Fortschritte	Fortschr.	Physik	Phys.
Freshwater	No abbrev.	Physiology	Physiol.
Gazette	Gaz.	Pollution	Pollut.
General	Gen.	Proceedings	Proc.
Genetics	Genet.	Psychological	Psychol.
Geographical	Geogr.	Publications	Publ.
Geological	Geol.	Quarterly	Q.
Geologische	Geol.	Rendus	R.
Gesellschaft	Ges.	Report	Rep.
Helvetica	Helv.	Research	Res.
History	Hist.	Review	Rev.
Immunity	Immun.	Revue, Revista	Rev.
Immunology	Immunol.	Rivista	Riv.
Industrial	Ind.	Royal	R.
Institute	Inst.	Science	Sci.
International	Int.	Scientific	Sci.
Jahrbuch	Jahrb.	Series	Ser.
Jahresberichte	Jahresber.	Service	Serv.
Japan, Japanese	Jpn.	Society	Soc.
Journal	J.	Special	Spec.
Laboratory	Lab.	Station	Stn.
Magazine	Mag.	Studies	Stud.
Ì 'aterial	Matr.	Surgery	Surg.
Mathematics	Math.	Survey	Surv.
Mechanical	Mech.	Symposia	Symp.
Medical	Med.	Symposium	Symp.
Medicine	Med.	Systematic	Syst.
Methods	No abbrev.	Technical	Tech.
Microbiological	Microbiol.	Technik	Tech.
Microbiology	Microbiol.	Technology	Technol.
Monographs	Monogr.	Therapeutics	Ther.
Monthly	Mon.	Transactions	Trans.
Morphology	Morphol.	Tropical	Trop.
National	Natl.	United States	U.S.
Natural, Nature	Nat.	University	Univ.
Neurology	Neurol.	Untersuchung	Unters.
Nuclear	Nucl.	Urological	Urol.
Nutrition	Nutr.	Verhandlungen	Verh.
Obstetrical	Obstet.	Veterinary	Vet.
Official	Off.	Virology	Virol.
Organic	Org.	Vitamin	Vitam.
Paleontology	Paleontol.	Wissenschaftliche	Wiss.
Pathology	Pathol.	Zeitschrift	Z.
Pharmacology	Pharmacol.	Zentralblatt	Zentralbl.
Philosophical	Philos.	Zoologie	Zool.
Physical	Phys.	Zoology	Zool.

APPENDIX 2

Abbreviations That May Be Used Without Definition in Table Headings (30)*

Term	Abbreviation
Amount	amt
Approximately	approx
Average	avg
Concentration	concn
Diameter	diam
Experiment	expt
Height	ht
Molecular weight	mol wt
Month	mo
Number	no.
Preparation	prepn
Specific activity	sp act
Specific gravity	sp gr
Temperature	temp
Trace	tr
Versus	vs
Volume	vol
Week	wk
Weight	wt
Year	yr

* In addition to the terms listed, abbreviations for units of measure are accepted without definition.

APPENDIX 3

Common Errors in Style and in Spelling

Wrong	Right
acetyl-glucosamine	acetylglucosamine
acid fast bacteria	acid-fast bacteria
acid fushsin	acid fuchsine
acridin orange	acridine orange
acriflavin	acriflavine
aesculin	esculin
airborn	airborne
air-flow	airflow
ampoul	ampoule
analagous	analogous
analize	analyze
bacteristatic	bacteriostatic
baker's yeast	bakers' yeast
baseline	base line (n.), base-line (adj.)
bi-monthly	bimonthly
bio-assay	bioassay
biurette	biuret
blendor	blender
blood sugar	blood glucose
bromcresol blue	bromocresol blue
by-pass	bypass
byproduct	by-product
can not	cannot
catabolic repression	catabolite repression
chloracetic	chloroacetic
clearcut	clear-cut
colicine	colicin
coverslip	cover slip
coworker	co-worker
cross over (n.)	crossover
crossover (v.)	cross over
darkfield	dark field
data is	data are
desoxy-	deoxy-
dessicator	desiccator

162

Wrong	Right
dialise	dialyze
disc	disk
Ehrlenmeyer flask	Erlenmeyer flask
electronmicrograph	electron micrograph
electrophorese	subject to electrophoresis
eukaryote	eucaryote
fermenter (apparatus)	fermentor
fermentor (organism)	fermenter
ferridoxin	ferredoxin
flourite	fluorite
fluorescent antibody technique	fluorescent-antibody technique
fungous (n.)	fungus
fungus (adj.)	fungous
gelatine	gelatin
germ-free	germfree
glucose-6-phosphate	glucose 6-phosphate
glycerin	glycerol
glycollate	glycolate
gonnorhea	gonorrhea
Gram-negative	gram-negative
gram stain	Gram stain
gyrotory	gyratory
halflife	half-life
haptene	hapten
Hela cells	HeLa cells
Hep-2-cells	HEp-2 cells
herpes virus	herpesvirus
hydrolize	hydrolyze
hydrolyzate	hydrolysate
immunofluorescent techniques	immunofluorescence techniques
india ink	India ink
indol	indole
innocula	inocula
iodimetric	iodometric
ion exchange resin	ion-exchange resin
isocitritase	isocitratase
keiselguhr	kieselguhr
large concentration	high concentration
less data	fewer data
leucocyte	leukocyte

Wrong	Right
little data	few data
low quantity	small quantity
mediums	media
melenin	melanin
merthiolate	Merthiolate
microphotograph	photomicrograph
mid-point	midpoint
moeity	moiety
much data	many data
new-born	newborn
occurrance	occurrence
over-all	overall
papergram	paper chromatogram
paraffine	paraffin
Petri dish	petri dish
phenolsulfophthalein	phenolsulfonephthalein
phosphorous (n.)	phosphorus
phosphorus (adj.)	phosphorous
pipet	pipette
planchette	planchet
plexiglass	Plexiglas
post-mortem	postmortem
prokaryote	procaryote
pyocine	pyocin
pyrex	Pyrex
radio-active	radioactive
regime	regimen
re-inoculate	reinoculate
saltwater	salt water
sea water	seawater
selfinoculate	self-inoculate
semi-complete	semicomplete
shelflife	shelf life
sidearm	side arm
small concentration	low concentration
spore-forming	sporeforming
stationary phase culture	stationary-phase culture
step-wise	stepwise

Wrong	Right
students' T test	Student's t test
sub-inhibitory	subinhibitory
T_2 phage	T2 phage
technic	technique
teflon	Teflon
thiamin	thiamine
thioglycollate	thioglycolate
thyroxin	thyroxine
transfered	transferred
transfering	transferring
transferrable	transferable
trichloracetic acid	trichloroacetic acid
tris-(hydroxymethyl)amino-methane	tris(hydroxymethyl)aminomethane
trypticase	Trypticase
tryptophane	tryptophan
ultra-sound	ultrasound
un-tested	untested
urinary infection	urinary tract infection
varying amounts of cloudiness	varying cloudiness
varying concentrations (5, 10, 15 mg/ml)	various concentrations (5, 10, 15 mg/ml)
waterbath	water bath
wave length	wavelength
X-ray (n.)	X ray
X ray (adj.)	X-ray
zero-hour	zero hour

APPENDIX 4

Words and Expressions to Avoid

Jargon	Preferred Usage
a considerable amount of	much
a majority of	most
a number of	many
absolutely essential	essential
accounted for by the fact	because
along the lines of	like
an order of magnitude faster	10 times faster
are of the same opinion	agree
as a consequence of	because
as a matter of fact	in fact (or leave out)
as is the case	as happens
as of this date	today
as to	about (or leave out)
at an earlier date	previously
at the present time	now
at this point in time	now
based on the fact that	because
by means of	by, with
completely full	full
consensus of opinion	consensus
definitely proved	proved
despite the fact that	although
due to the fact that	because
during the course of	during, while
elucidate	explain
end result	result
entirely eliminate	eliminate
fabricate	make
fewer in number	fewer
finalize	end
first of all	first
for the purpose of	for
for the reason that	since, because
from the point of view of	for

Jargon	Preferred Usage
give rise to	cause
has the capability of	can
having regard to	about
in a number of cases	some
in a position to	can, may
in a satisfactory manner	satisfactorily
in a very real sense	in a sense (or leave out)
in case	if
in close proximity	close, near
in connection with	about, concerning
in many cases	often
in my opinion it is not an un- justifiable assumption that	I think
in order to	to
in relation to	toward, to
in respect to	about
in some cases	sometimes
in terms of	about
in the event that	if
in the possession of	has, have
in view of the fact that	because, since
inasmuch as	for, as
initiate	begin, start
is defined as	is
it has been reported by Smith	Smith reported
it has long been known that	I haven't bothered to look up the reference
it is apparent that	apparently
it is believed that	I think
it is clear that	clearly
it is clear that much additional work will be required before a complete understanding	I don't understand it
it is doubtful that	possibly
it is evident that *a* produced *b*	*a* produced *b*
it is of interest to note that	(leave out)
it is often the case that	often
it is suggested that	I think
it is worth pointing out in this context that	note that

Jargon	Preferred Usage
it may be that	I think
it may, however, be noted that	but
it should be noted that	note that (or leave out)
it was observed in the course of the experiments that	we observed
lacked the ability to	couldn't
large in size	large
let me make one thing perfectly clear	a snow job is coming
militate against	prohibit
needless to say	(leave out, and consider leaving out whatever follows it)
of great theoretical and practical importance	useful
on a daily basis	daily
on account of	because
on behalf of	for
on the basis of	by
on the grounds that	since, because
on the part of	by, among, for
our attention has been called to the fact that	we belatedly discovered
owing to the fact that	since, because
perform	do
pooled together	pooled
prior to	before
protein determinations were performed	proteins were determined
quite unique	unique
rather interesting	interesting
red in color	red
referred to as	called
relative to	about
resultant effect	result
smaller in size	smaller
subsequent to	after
sufficient	enough
take into consideration	consider
terminate	end
the great majority of	most

Jargon	Preferred Usage
the opinion is advanced that	I think
the question as to whether	whether
the reason is because	because
there is reason to believe	I think
this result would seem to indicate	this result indicates
through the use of	by, with
ultimate	last
utilize	use
was of the opinion that	believed
ways and means	ways, means (not both)
we have insufficient knowledge	we don't know
we wish to thank	we thank
with a view to	to
with reference to	about (or leave out)
with regard to	concerning, about (or leave out)
with respect to	about
with the possible exception of	except
with the result that	so that

Sermons on brevity and chastity are about equally effective.
Verbal promiscuity flows from poverty of language and obe-
sity of thought, and from an unseemly haste to reach print—
a premature ejaculation, as it were.

—ELI CHERNIN

APPENDIX 5

Prefixes and Abbreviations for SI (Système International) Units

No.	Prefix	Abbreviation
10^{-18}	atto	a
10^{-15}	femto	f
10^{-12}	pico	p
10^{-9}	nano	n
10^{-6}	micro	μ
10^{-3}	milli	m
10^{-2}	centi	c
10^{-1}	deci	d
10	deka	da
10^2	hecto	h
10^3	kilo	k
10^6	mega	M
10^9	giga	G
10^{12}	tera	T
10^{15}	peta	P
10^{18}	exa	E

APPENDIX 6

Accepted Abbreviations and Symbols

Term	Abbreviation or Symbol	Term	Abbreviation or Symbol
absorbance	A	flavin adenine dinucleotide	FAD
acetyl	Ac	flavin mononucleotide	FMN
adenine	Ade	fouled up (beyond all rec-	FUBAR
adenosine 5'-diphosphate	ADP	ognition)	
adenosine 5'-monophos-	AMP	gram	g
phate		guanidine	Gdn
adenosine 5'-triphosphate	ATP	guanine	Gua
adenosine triphosphatase	ATPase	guanosine	Guo
alanine	Ala	guanosine 5'-diphosphate	GDP
alternating current	a.c.	hemoglobin	Hb
ampere	A	hemoglobin, oxygenated	HbO_2
antibody	Ab	henry	H
antigen	Ag	heptyl	Hp
arabinose	Ara	hertz	Hz
bacille Calmette-Guérin	BCG	hexyl	Hx
biological oxygen demand	BOD	horsepower	hp
blood urea nitrogen	BUN	hour	h
boiling point	bp	infrared	IR
candela	cd	inosine 5'-diphosphate	IDP
central nervous system	CNS	intravenous	i.v.
coenzyme A	CoA	isoleucyl	Ile
coulomb	C	joule	J
cytidine	Cyd	kelvin	K
cytidine 5'-diphosphate	CDP	kilogram	kg
cytidine 5'-monophos-	CMP	kinetic energy	KE
phate		lethal dose, median	LD_{50}
cytidine 5'-triphosphate	CTP	leucyl	Leu
cytosine	Cyt	litre (liter)	l
deoxyribonuclease	DNase	lysinyl	Lys
deoxyribonucleic acid	DNA	melting point	mp
deoxyuridine monophos-	dUMP	messenger ribonucleic acid	mRNA
phate		meta-	m-
diethylaminoethyl cellulose	DEAE-cellu-	methionyl	Met
	lose	methyl	Me
electrocardiogram	ECG	metre (meter)	m
electroencephalogram	EEG	milliequivalent	meq
ethyl	Et	minimum lethal dose	MLD
ethylenediamine-	EDTA	minute (time)	min
tetraacetate		molar (concentration)	M
farad	F	mole	mol

171

Term	Abbreviation or Symbol	Term	Abbreviation or Symbol
muramic acid	Mur	seryl	Ser
newton	N	siemens	S
nicotinamide adenine dinucleotide	NAD	species	sp. (sing.), spp. (pl.)
nicotinamide adenine dinucleotide (reduced)	NADH	standard deviation	SD
		standard error	SE
normal (concentration)	*N*	standard temperature and pressure	STP
nuclear magnetic resonance	NMR	subcutaneous	s.c.
ohm	Ω	tobacco mosaic virus	TMV
ornithyl	Orn	tonne (metric ton)	t
ortho-	*o-*	transfer ribonucleic acid	tRNA
orthophosphate	P_i	tris(hydroxymethyl)aminomethane	Tris
outside diameter	o.d.		
para-	*p-*	tyrosinyl	Tyr
pascal	Pa	ultraviolet	UV
phenyl	Ph	*United States Pharmacopeia*	USP
plaque-forming units	PFU		
purine	Pur	uracil	Ura
pyrophosphate	PP_i	uridine 5'-diphosphate	UDP
radian	rad	volt	V
reticuloendothelial system	RES	volume	*V*
ribonuclease	RNase	watt	W
ribonucleic acid	RNA	week	wk
ribose	Rib	white blood cells (leukocytes)	WBC
ribosomal ribonucleic acid	rRNA		
roentgen	R	xanthine	Xan
second (time)	s	xanthosine	Xao
serum glutamic oxalacetic transaminase	SGOT	xanthosine 5'-diphosphate	XDP
		year	yr

References

1. **Aaronson, S.** 1977. Style in scientific writing. Current Contents, No. 2, 10 January; p. 6-15.
2. **American National Standards Institute, Inc.** 1969. American national standard for the abbreviation of titles of periodicals. ANSI Z39.5-1969. American National Standards Institute, Inc., New York.
3. **American National Standards Institute, Inc.** 1979. American national standard for writing abstracts. ANSI Z39.14-1979. American National Standards Institute, Inc., New York.
4. **American National Standards Institute, Inc.** 1972. American national standard for the preparation of scientific papers for written or oral presentation. ANSI Z39.16-1972. American National Standards Institute, Inc., New York.
5. **American National Standards Institute, Inc.** 1977. American national standard for synoptics. ANSI Z39.34-1977. American National Standards Institute, Inc., New York.
6. **Anderson, J. A., and M. W. Thistle.** 1947. On writing scientific papers. Bull. Can. J. Res., 31 December 1947, N.R.C. No. 1691.
7. **Barnes, G. A.** 1982. Communication skills for the foreign-born professional. ISI Press, Philadelphia.
8. **Bernstein, T. M.** 1965. The careful writer. A modern guide to English usage. Atheneum, New York.
9. **Booth, V.** 1981. Writing a scientific paper and speaking at scientific meetings. 5th ed. The Biochemical Society, London.
10. **Burch, G. E.** 1954. Of publishing scientific papers. Grune & Stratton, New York.
11. **CBE Style Manual Committee.** 1978. Council of Biology Editors style manual: a guide for authors, editors, and publishers in the biological sciences, 4th ed. Council of Biology Editors, Inc., Washington, D.C.
12. **Chase, S.** 1953. Power of words. Harcourt, Brace and Co., New York.
13. **Council of Biology Editors.** 1968. Proposed definition of a primary publication. Newsletter, Council of Biology Editors, November 1968, p. 1-2.
14. **Cremmins, E. T.** 1982. The art of abstracting. ISI Press, Philadelphia.
15. **Day, R. A.** 1973. Economics of printing, p. 45-56, *In* R. A. Day et al. (ed.), Economics of scientific publications. Council of Biology Editors, Washington, D.C.
16. **Day, R. A.** 1975. How to write a scientific paper. ASM News **41**:486-494.
17. **Day, R. A.** 1976. Scientific journals: an endangered species. ASM News **42**:288-291.
18. **DeBakey, L.** 1976. The scientific journal. Editorial policies and practices. Guidelines for editors, reviewers, and authors. The C. V. Mosby Co., St. Louis.
19. **Garfield, E.** 1977. Essays of an information scientist, vol. 1, p. 10. ISI Press, Philadelphia.
20. **Houghton, B.** 1975. Scientific periodicals; their historical development, characteristics and control. Linnet Books, Hamden, Conn.
21. **Huth, E. J.** 1982. How to write and publish papers in the medical sciences. ISI Press, Philadelphia.
22. **International Committee of Medical Journal Editors.** 1982. Uniform requirements for manuscripts submitted to biomedical journals. Br. Med. J. **284**:1766-1770 [and Ann. Intern. Med. **96**:766-770].
23. **McGirr, C. J.** 1978. Guidelines for abstracting. Tech. Commun. **25**(2):2-5.
24. **Meyer, R. E.** 1977. Reports full of "gobbledygook." J. Irreproducible Res. **22**(4):12.
25. **Michaelson, H. B.** 1982. How to write and publish engineering papers and reports. ISI Press, Philadelphia.
26. **Mitchell, J. H.** 1968. Writing for professional and technical journals. John Wiley & Sons, Inc., New York.
27. **Morrison, J. A.** 1980. Scientists and the scientific literature. Scholarly Publishing **11**:157-167, 1980.

28. **O'Connor, M., and F. P. Woodford.** 1975. Writing scientific papers in English. Associated Scientific Publishers, Amsterdam.
29. **Peterson, M. S.** 1961. Scientific thinking and scientific writing. Reinhold Publishing Corp., New York.
30. **Pollock, G.** 1977. Handbook for ASM editors. American Society for Microbiology, Washington, D.C.
31. **Ratnoff, O. D.** 1981. How to read a paper. *In* K. S. Warren (ed.), Coping with the biomedical literature, p. 95–101. Praeger, New York.
32. **Reid, W. M.** 1978. Will the future generations of biologists write a dissertation? BioScience **28:**651–654.
33. **Rosten, L.** 1968. The joys of Yiddish. McGraw-Hill Book Co., New York.
34. **Sagan, C.** 1977. The dragons of Eden. Speculations on the evolution of human intelligence. Ballantine Books, New York.
35. **Siegenthaler, W., and R. Lüthy (ed.).** 1978. Current chemotherapy. Proceedings of the 10th International Congress of Chemotherapy. American Society for Microbiology, Washington, D.C., 2 vol.
36. **Skillin, M. E., R. M. Gay, et al.** 1974. Words into type, 3rd ed. Prentice-Hall, Inc., Englewood Cliffs, N.J.
37. **Strunk, W., Jr., and E. B. White.** 1979. The elements of style. Third Edition. The Macmillan Co., New York.
38. **Trelease, S. F.** 1958. How to write scientific and technical papers. The Williams & Wilkins Co., Baltimore.
39. **Weiss, E. H.** 1982. The writing system for engineers and scientists. Prentice-Hall, Inc., Englewood Cliffs, N.J.
40. **Woodford, F. P. (ed.).** 1968. Scientific writing for graduate students. A Council of Biology Editors Manual. The Rockefeller University Press, New York.
41. **Zinsser, W.** 1976. On writing well. An informal guide to writing nonfiction. Harper & Row, Publishers, New York.

Index